Math Remediation for the College Bound

How Teachers Can Close the Gap, from the Basics through Algebra

Daryao S. Khatri and Anne O. Hughes

Rowman & Littlefield Education
A division of
ROWMAN & LITTLEFIELD PUBLISHERS, INC.
Lanham • New York • Toronto • Plymouth, UK

Published by Rowman & Littlefield Education
A division of Rowman & Littlefield Publishers, Inc.
A wholly owned subsidiary of The Rowman & Littlefield Publishing Group, Inc.
4501 Forbes Boulevard, Suite 200, Lanham, Maryland 20706
http://www.rowmaneducation.com

Estover Road, Plymouth PL6 7PY, United Kingdom

Copyright © 2012 by Daryao S. Khatri and Anne O. Hughes

All rights reserved. No part of this book may be reproduced in any form or by any electronic or mechanical means, including information storage and retrieval systems, without written permission from the publisher, except by a reviewer who may quote passages in a review.

British Library Cataloguing in Publication Information Available

Library of Congress Cataloging-in-Publication Data

Khatri, Daryao S., 1945-
 Math remediation for the college bound. How teachers can close the gap, from the basics through algebra / Daryao Khatri, Anne O. Hughes.
 p. cm.
  ISBN 978-1-61048-366-7 (hardback) — ISBN 978-1-61048-367-4 (paper) — ISBN 978-1-61048-368-1 (electronic)
 1. Algebra. I. Hughes, Anne O. II. Title.
 QA152.3.K47 2011
 512—dc22   2011002316

∞™ The paper used in this publication meets the minimum requirements of American National Standard for Information Sciences—Permanence of Paper for Printed Library Materials, ANSI/NISO Z39.48-1992. Printed in the United States of America

Contents

Preface		v
Introduction		ix
Foreword, William L. Pollard		xiii
1	The Gateway Academic Program: Five Teaching Strategies for Basic Math and Algebra	1
2	Finding Gaps and Honing Basic Skills	23
3	Mental Math	45
4	Integers and Arithmetic Operations	55
5	Data, Statistics, and Graphs	81
6	A Sometimes Love–Hate Relationship: Facts about Triangles, Intersecting Lines, and Units of Measurements	95
7	The Two Great Devils: The Negative Sign and Negative Numbers	107
8	Change for Currency Notes: Simple Fractions	115
9	Exponents: Powers of 10	123
10	Powers of Variables and Constants	137
11	The World of Expressions and Elementary Equations	151
12	The Algebra Savior: Cross-Products	163

Introduction

The immediate purpose of this book is to assist college faculty in closing the gaps in basic math and algebra being experienced by entering college freshmen who have found that they must take remedial math courses and by graduating high school students who have fallen behind in their math preparation required for math courses in college. The long-range purpose of the book is to eliminate the need for remedial math education at the college level.

The unique features of this book are:

- Experienced teachers can carefully select math topics and experiences familiar to all of the students as the bases for building necessary concepts and skills, thereby saving time and creating vivid examples that are more easily mastered and retained over time.
- The only graphics included are those required for the solution of a problem.
- The book is designed for a "calculator-free environment." So, instead of students' fingers clicking on the calculator's keys, their minds become the calculators.
- Needlessly difficult and tedious problems are avoided. Instead, the emphasis is on pattern recognition and mastery of required facts and operations.
- Decimals are used only when teaching them in relation to fractions, percentages, proportions, and money.
- The topics are smoothly sequenced so that each successive one builds upon those topics that have come before. Depending on the students'

gaps, we expect teachers to pick and choose topics needed for their students' academic preparations.
- Once the basic concepts and processes are clearly understood, they are practiced until they are completely mastered. For example, we have discovered that many of our students haven't fully learned the multiplication tables 1 through 9, let alone through 15. This gap slows down their problem solving on virtually every task and dilutes their self-confidence. Therefore, this book provides frequent and varied reviews of everything being taught, including the multiplication tables.
- The same teaching methods that have been unsuccessful with students in previous math courses are not likely to be successful with these same students this time around. So, the two beginning chapters of the book emphasize teaching and management strategies that have proven effective with students over the years and have kept the retention and passing rates in our classes and the summer programs at close to 100 percent. All of the math presentations and practice in the chapters that follow are designed and sequenced with these teaching and management strategies in mind.

Given these general characteristics of the book as a whole, the content of each chapter is briefly described below. Chapter 1 defines and illustrates the five teaching strategies and their respective skill clusters that have been developed and successfully tested over the years. Chapter 2 presents several days of suggested instructional activities to identify rapidly the students' math gaps.

Beginning with chapter 3, "Mental Math," the focus shifts to the math content and continues so for the remainder of the book. Each of these succeeding chapters provides clear explanations and examples of the topics and operations involved, practice exercises, and tests, as well as extensive reviews of what has been covered in preceding chapters. For example, in the basic mathematics chapters, chapters 3 through 9, the multiplication tables are continually reviewed, because they are a major key to the mastery evidenced in quick mental problem solving.

In addition, some of the topics included in the basic mathematics chapters often are not presented until the very end of typical algebra textbooks. A case in point is statistics. In this book, it comes early, appearing as chapter 5, "Data, Statistics, and Graphs." Statistical presentations in one form or another are everywhere—in sports, politics, history, business, education, the sciences, and more. Students are familiar with them in some form. Statistics also uses terminology, symbols, graphical forms, and equation formats drawn from algebra. The chapter is a way of presenting some new material related to algebra using the students' knowledge.

Chapter 9, "Exponents: Powers of 10," concludes the basic math part of the book, but the students will see "old friends" from this and the other chapters as they transition into algebra. They still will have to apply the four basic arithmetical operations, use the multiplication tables, remember the rules about 0 and 1 and negative numbers, handle fractions, deal with exponents, and so on.

Without the students having complete mastery of the topics and processes included in these early chapters, the bridge into algebra will be a rickety one. Our primary teaching strategy is inductive, that is, moving from the familiar (known) to the unfamiliar or possibly forgotten (unknown). In addition, every one of the students should experience success right from the start and develop the expectation of continuing to be successful. Not a single student is to be left behind.

With chapter 10, "Powers of Variables and Constants," the bridge into algebra begins with definitions, operations, and practice involving variables and constants. With chapter 11, "The World of Expressions and Elementary Equations," the bridge into algebra has been crossed.

Chapters 12 through 20 focus entirely on algebra and its applications. Again, the inductive strategy of moving from the familiar to the unfamiliar is applied. It is particularly apparent in chapters 13, "Ratio and Proportion," and 14, "Just Shopping for the Best Deal: Discount, Sales Tax, Commission, Percentage, and Simple Interest." We might also point out that the topics of chapters 18–20 are very popular with the designers of math placement tests at the college entrance level.

With regard to mental exercises in the book, teachers are reminded that they need to review problems contained under these sections on a regular basis until all students are proficient with them. In their daily reviews, the teachers need to pay special attention to students who are still having difficulties with mental problems; just keep them engaged as soon as you begin review. Also, keep them engaged throughout the class by directing questions to them based on their differential knowledge base.

lay down the "rules of the road" for the course or program, gain the students' acceptance of them as being in their best interest, and then enforce them with the help of the students.

We use an outline form for the presentation of the management strategy and its clusters of skills because it is probably the most efficient way to present them, and they do not need the explanations and examples required for the other teaching strategies and their skills. However, we recommend that you, as the teacher, pass the listing out to your students and then go over it with them. That is why the student part of the listing is specifically addressed to them.

Clusters of Skills for Successful Student Performance

1. Time and attendance
    1.1. Be in class on time.
    1.2. Always come to class.
    1.3. Patronize the restrooms before coming to class—never during a presentation unless you have an emergency.
    1.4. If you are late or do have to miss a class, before coming to the next class, copy the notes of a colleague from the missed class, obtain a copy of the previous day's agenda for your notebook, and study the missed material.

2. Organization of Academic Resources
    2.1. A three-ring 1½-inch D-ring binder for daily note-taking and handouts
    2.2. Index tabs for the number of weeks of the course or program
    2.3. Loose sheets of paper punched for a three-ring binder
    2.4. Pencils and pens
    2.5. Forget the calculator—you won't be needing it
    (The requirement for these resources is more important than it may seem on the surface. Here's why: This requirement forces you as a student to organize your work efficiently, and it allows your teacher to check on your note-taking skills easily by matching what you have taken down compared to the teacher's own notebook. If some of you are not doing an adequate job, then your teacher may need to schedule a special note-taking and organization session for you and other students who may be having difficulties as well. The arts of good organization and note-taking are critical to success in a course.)

3. Note-Taking
    3.1. Take down the notes as presented by the teacher on the chalkboard, screen, or flip chart.

  3.2. Expect the teacher to check your notes to make certain you are doing them and they are accurate.
  3.3. Date your notes.
  3.4. In the case of absence, borrow the notes of another student. (You are responsible, not the teacher.)
4. Classroom Learning Etiquette
  4.1. Follow the rule of *no* cell phones, iPods, or pagers in the classroom. They are distractions to your learning.
  4.2. Feel free to raise your hand, but *wait to be recognized* by the teacher before either asking or answering a question or making a comment.
  4.3. Avoid talking with one another during the class session unless the teacher sanctions it.
  4.4. Avoid closing your eyes or putting feet on the table or chair.
  4.5. Stay awake—no sleeping.
  4.6. Save personal grooming for the restroom—not the classroom.
  4.7. Avoid working on another course—it's bad manners, and it shows you're not very organized.
  4.8. Avoid such common student "cop-out" expressions as "I'm bored," "I don't understand," "I'm confused," and "Would you repeat that, please?"
  4.9. Agree to no eating or drinking.
5. Homework
  5.1. Have your homework ready for checking as soon as you arrive in your class.
  5.2. Have straight practice homework completed by the next class session.
  5.3. Complete priming homework, if possible; if not, be ready with your questions.

Cluster of Skills for Successful Teacher Performance

1. Learn the students' names as fast as possible. Being able to call the students by name is absolutely essential. It tells the students that you care about them as individuals, and it is a critical tool of classroom control and the effective use of the teaching strategies. We reiterate: *If you do not know the names of students, you will not be able to use either the management or the teaching strategies effectively*.
2. Go over the classroom rules with the students, and make sure they understand why they are so important. Then have the students sign a pledge to abide by the rules.

3. Be consistent in enforcing the "rules" outlined above for the students. The first week or two will be the hardest, but the students will begin to see the payoff very soon.
4. Have all of your needed materials sequentially organized for the class so you will not waste time looking for the next task or activity.
5. Pass out the agenda, and go over its topics, before starting the lesson for that day. (The agenda lets the students know what to expect, and it keeps you on track as well.)
6. Keep presentations as short as possible; avoid unnecessary elaborations or illustrations. (You want the students to become involved.)
7. Assign reasonable homework for practice. Students should be able to finish the homework in an hour or so. (Sometimes, just copying a complex problem a few times really assists both in understanding it and then in mastering it.)
8. Check practice homework quickly for each student at the beginning of the following class session. If a student has not done it, find out why, and make arrangements for the student to complete it before leaving for the day, if possible. Make arrangements for the homework to be completed in a timely fashion. Do not let it slide.
9. Check priming homework, and briefly go through the skills and information needed to solve the problem, if needed.
10. Always stand when making a presentation or interacting with students. (Standing, you are taller and more commanding than any one of the seated students. And remember: you are not one of the gang; you are the leader of the group.)
11. Avoid "put-down" comments, such as "I just covered that," or "You should have known."
12. Follow the organization of your own academic resources by using the same kind of three-ring binder, tabs, and so forth you are requiring of the students. Let the students examine your binder if they want to see how it works. Be sure to compliment students on their notebooks if they have done a terrific job with them.
13. Inform the students they are *not* to open the course or program textbook—if there is one—during your presentations.
14. Encourage the students to be in contact with one another about their homework assignments or preparing for quizzes and exams outside of class time. Print out a list of their names, telephone or cell phone numbers, and e-mail addresses.

INDUCTIVE TEACHING STRATEGY AND ITS SKILL CLUSTERS

Of the five major strategies we have identified, we have concluded that the inductive teaching strategy is probably the most critical one. Of course, we must confess that Socrates arrived at this truth first, and he was a master of it.

In essence, our version of the inductive strategy requires the teacher to elicit the principles, answers, and insights from the students themselves. That is, the teacher does not tell them the specific answers to a problem or define the concepts, but instead leads the students to see or formulate what they are through the use of a cluster of teaching skills.

The skills of the inductive strategy are: anchoring, analogical reasoning, students' differential knowledge bases as identified by the teacher, use of this knowledge by the teacher to guarantee equal student participation, and parallel organization and teaching. A good teacher will typically employ all of these skills in the course of a single class session. The skill clusters of the inductive strategy are next described and illustrated.

Anchoring

In a class session involving the inductive teaching strategy, the teacher begins with the anchoring skill. *Anchoring* is using information or a topic that is completely familiar to all of the students—the known (e.g., a half dollar or a quarter, representing fifty and twenty-five cents, respectively). Then the teacher uses the familiar example(s) to introduce new information, abstract concepts, and/or topics to the students—the unknown—such as fractional, decimal, and percent notations.

Naturally, the first step is to make certain the students are absolutely familiar with whatever the concrete (known) examples are—in this case, the American monetary system. In making certain that the students really are familiar with the topic chosen to begin the process, you are engaging in the skill of anchoring: *selecting a familiar topic in the students' experience and using it to lead them into seeing the connection(s) to the new or less familiar concepts or topic(s).*

Analogical Reasoning

Another key point to remember about the inductive strategy is that it involves recognizing that some characteristic is similar between the known particulars (currency value in cents) and an unknown concept (fractional notation). In this instance, the fractional notation is obtained by dividing the currency value expressed in cents (by 100), followed by simplifying the fraction. This recognition of a key common characteristic among seemingly dissimilar entities is known as *analogical thinking* or *reasoning by analogy*.

A skilled teacher can lead students to see the similarity or analogy. Additionally, on the surface, 0.5 and ½ may look dissimilar, but the correct inference is that both can represent fifty cents and are half of something.

Let us illustrate the inductive strategy and its skills cluster in action using two examples.

Inductive Strategy in Basic Math: Using Anchoring, Students' Differential Knowledge Bases, Equal Participation, and Analogical Reasoning—Example 1

First, create a table with five columns, nine rows, and currency value in the second column as shown in table 1.1. The row numbers are in the first column. This example uses students' prior knowledge with currency (an example of anchoring, as described above).

Table 1.1 An Incomplete Arrangement of Fractional, Percent, and Decimal Notations

Item #	Currency Value (Known)	Fraction Notation (Unknown?)	Percent Notation (Unknown?)	Decimal Notation (Unknown?)
1	50 cents or half a dollar			
2	25 cents or a quarter of a dollar			
3	10 cents			
4	100 cents or one full dollar			
5	75 cents			
6	5 cents			
7	9 cents			
8	13 cents			
9	1 cent			

Engage students in filling the rest of the information in rows first, and be sure to call each student by name. Avoid starting with one dollar, because some students may have problems in understanding that it is actually 100 percent. In filling in the rows in the table, we use another one of the key skills in the inductive strategy: the *students' differential knowledge bases*. Briefly defined, this fancy term describes the fact that no two students possess exactly the same knowledge about a given subject. Some will always know more than others—and there usually is quite a range.

Some students will know more about one topic than another topic. As a teacher, you need to find out very quickly what the individual knowledge bases are—and then tailor your questions to fit students' individual knowledge levels. The idea is for every student to contribute positively to the learning process and thereby to succeed.

1. Start with a student less prepared in math; be sure to call the student by name. Hold up a half dollar and ask, "Mark, what do I have here?"

The student should readily identify that the coin is a half dollar or fifty-cent piece. Then ask the student if the half dollar, which is half of a dollar, might be written in another way to show half of something else and point to either the percent or fraction column. If this student has problems completing the information in row 1, you may have to ask several students to assist in completing it.
2. Ask another less prepared student to help you fill in the information for the second row. Repeat this approach for the third row.
3. Now ask a slightly better prepared student to help you fill in row 4.
4. Go around the room, asking other students to help you fill in row 5.
5. Ask a better prepared student to fill in information for row 6.
6. Ask some other students to help you fill in information for rows 7 and 8.
7. Start completing information for row 9 with less prepared students. If the answers provided are not correct, ask some well-prepared students to help you complete the information for this row.

In moving from student to student and calling each one by name, you, as the teacher, are also employing the skill of *equal participation*. That is, by the time you finish eliciting the information from the students, virtually every student in the class will have contributed successfully to the completion of the task. It won't just be the three eager "swifties" in the front row. Table 1.2 on the next page shows the completed grid.

Now, ask several students to draw conclusions about changing from decimal notation to percent notation and vice versa (from percent notation to decimal notation). On the board, write their concluding statements in the following ways:

Decimal Notation → Percent Notation
(Multiply by 100)

Percent Notation → Decimal Notation
(Divide by 100)

When the table has been completely filled in and the two conclusions have been stated, let students know that the class will repeat these rules, with illustrations daily for changing from one notation to the other and examples as needed, until the new knowledge is fully mastered and habituated. Make certain the students have recorded these concluding statements in their notebooks, as well as the tables if they've filled out individual copies. Also, involve the initially not-so-well-prepared students in answering some of the more challenging questions as their mastery improves.

Table 1.2 A Completed Example of Fraction, Percent, and Decimal Notations

Item #	Currency Value	Fraction Notation	Percent Notation	Decimal Notation
1	50 cents	$\frac{50}{100} = \frac{1}{2}$	50%	0.5
2	25 cents	$\frac{25}{100} = \frac{1}{4}$	25%	0.25
3	10 cents	$\frac{10}{100} = \frac{1}{10}$	10%	0.10
4	One dollar (100 cents)	$\frac{100}{100} = 1$	100%	1.0
5	75 cents	$\frac{75}{100} = \frac{3}{4}$	75%	0.75
6	5 cents	$\frac{5}{100} = \frac{1}{20}$	5%	0.05
7	9 cents	$\frac{9}{100}$	9%	0.09
8	13 cents	$\frac{13}{100}$	13%	0.13
9	1 cent	$\frac{1}{100}$	1%	0.01

As a teacher, you have undoubtedly observed from this example that many of the skills of the inductive strategy have been used in a seamless way. There is really no way to partition them. If a class session is to be successful, then you, as the teacher, must be prepared to move from one skill to another in working effectively with your students. And always you must be in control of the situation.

Here is another example.

Inductive Strategy for Introductory Algebra: Using Review, Anchoring, Students' Differential Knowledge Bases, and Equal Participation—Example 2

Review, with students' help, the powers of 10 in product and quotient forms. Involve as many students as possible. This review can be done either orally, on the board, or by using a combination approach (some problems on the board and others orally). The inductive review is presented in table 1.3. First, start writing 10, 100, and 1000 in powers of 10—for example:

$1000 = (10)(10)(10) = 10^3$
$100 = (10)(10) = 10^2$
$10 = (10) = 10^1$
$1 = 10^0$
$1,000,000 = (10^3)(10^3) = 10^6$

Table 1.3   Completed Presentations for Powers of 10

Item #	Algebraic Expression with Powers of 10	Details in Solving	Final Answer
1	$10^3 \cdot 10^5$	10^{3+5}	10^8
2	$10^{-3} \cdot 10^5$	10^{-3+5}	10^2
3	$10^3 \cdot 10^{-5}$	10^{3-5}	10^{-2}
4	$10^{-3} \cdot 10^{-5}$	10^{-3-5}	10^{-8}
5	$\dfrac{10^7}{10^3}$	10^{7-3}	10^4
6	$\dfrac{10^7}{10^{-3}}$	10^{7+3}	10^{10}
7	$\dfrac{10^{-7}}{10^{-3}}$	10^{-7+3}	10^{-4}
8	$\dfrac{10^{-7}}{10^3}$	10^{-7-3}	10^{-10}

Next, create table 1.4 as shown:

1. Start with a less prepared student in filling in the first several rows.
2. Based on students' differential knowledge bases, engage them in filling in the rest of the information in rows first. Start with less prepared students and ask those students to help you fill in the information in rows 1 through 4. You might have to ask several students to help in completing these rows.
3. Ask another somewhat better prepared student to help you fill in the information for the fifth row.
4. Now ask well-prepared students to help you fill in rows 6 through 8. Then ask other students to help you fill in rows 9 through 12.

5. Ask several very well-prepared students to help you fill in information for the last two rows (13 and 14).

Table 1.4 Algebraic Expressions of Exponents, an Incomplete Table

Item #	Algebraic Expression with Exponentials	Details in Solving	Final Answer
1	$x^3 \cdot x^5$		
2	$x^{-3} \cdot x^5$		
3	$x^3 \cdot x^{-5}$		
4	$x^{-3} \cdot x^{-5}$		
5	$\dfrac{x^7}{x^3}$		
6	$\dfrac{x^{-7}}{x^3}$		
7	$\dfrac{x^7}{x^{-3}}$		
8	$\dfrac{x^{-7}}{x^{-3}}$		
9	$(m^3)^3$		
10	$(m^{-4})^3$		
11	$(m^3)^{-5}$		
12	$(m^{-4})^{-2}$		
13	$(m^4)^{\frac{1}{2}}$		
14	$(z^3)^{\frac{1}{3}}$		

A complete table with final answers is shown in table 1.5.

Now with the help of a number of students, develop the general formulas for exponentials, as follows:

$$10^m \cdot 10^n = 10^{m+n} \text{ (Product Rule)}$$

$$\frac{p^m}{p^n} = p^{m-n} \text{ (Quotient Rule)}$$

$$(z^m)^n = z^{mn} \text{ (Power-Raised-to-a-Power Rule)}$$

Table 1.5 A Completed Presentation of Algebraic Expressions with Exponents

Item #	Algebraic Expression with Exponentials	Details in Solving	Final Answer
1	$x^3 \cdot x^5$	x^{3+5}	x^8
2	$x^{-3} \cdot x^5$	x^{-3+5}	x^2
3	$x^3 \cdot x^{-5}$	x^{3-5}	x^{-2}
4	$x^{-3} \cdot x^{-5}$	x^{-3-5}	x^{-8}
5	$\dfrac{x^7}{x^3}$	x^{7-3}	x^4
6	$\dfrac{x^7}{x^{-3}}$	x^{7+3}	x^x
7	$\dfrac{x^{-7}}{x^{-3}}$	x^{-7+3}	x^{-4}
8	$\dfrac{x^{-7}}{x^3}$	x^{-7-3}	x^{-10}
9	$(m^3)^3$	m^{3*3}	m^9
10	$(m^{-4})^3$	m^{-4*3}	m^{-12}
11	$(m^3)^{-5}$	m^{3*-5}	m^{-15}
12	$(m^{-4})^{-2}$	m^{-4*-2}	m^8
13	$(m^4)^{\frac{1}{2}}$	$m^{4*(1/2)}$	m^2
14	$(z^3)^{\frac{1}{3}}$	$z^{3*(1/3)}$	z

Make sure the students record the needed information in their notebooks. Finally, ask the students to describe the thinking processes they went through to arrive at their answers for this algebraic exercise. The more the students are in touch with the thinking techniques they are using with success, the more confidence they are likely to acquire in their ability to succeed—and the more likely they are to succeed.

Because of its importance to the inductive strategy and to successful teaching in general, we return to the skill of the students' differential knowledge bases. As we have noted before, the individual students in any class vary in terms of the amount of knowledge they can apply to the particular subject matter involved. Those with more knowledge should be given the more demanding questions to answer in the class sessions, while those with less are given the more basic questions to answer.

But here, we add a new dimension to it. That is, as your course progresses, you should gradually demand more from *all* of the students, so the less prepared ones can see their progress in the material and begin to lessen the gap in comparison with their other peers, while the more adequately prepared will continue to gain in their understandings. In short, we want you to raise the bar for every student's performance as the course or program progresses.

Importantly, not only will individual students gain confidence in their ability to succeed in learning the subject matter, but the entire class's confidence in their collective ability to learn will increase as well. As the old saying goes, "Nothing succeeds like success."

Two illustrations of the use of this skill are provided below—one from basic math and the other from introductory algebra.

Illustration of Students' Differential Knowledge Bases—Example 1: Using Squares and Square Roots in Basic Math

You are checking the students' instant recall of the squares and square roots of more commonly used whole numbers at the beginning of the course. Here are some applications of the skill for this topic.

1. Ask a less well prepared student by name for the square of 6: "Jay, what is the square of 6?" Then proceed through 10 with other similarly prepared students, always calling each one by name. Be sure to use each student's name, or others may call out. For each correct answer, say "Good" or "Great" or "That's right" or words to that effect.
2. Then, using the same basic approach, ask slightly more prepared students for the squares of 11 through 15. Then you can ask the better prepared students for the square roots of 144, 169, 225, and so on. After a few days of a class, start asking all students these facts about squares and square roots.
3. Another tactic you can employ here is to ask a well prepared student a difficult question. For example, ask this student for the cube root of 64. If the student answers correctly with 4, then ask a less well-prepared student to repeat the answer to the same question and have that particular student explain it. This practice of repetition reinforces the information and will also help to prevent students from drifting off the task because they do not know who will be called upon next.

Illustration of Students' Differential Knowledge Bases—Example 2: Finding the LCM and GCF in Introductory Algebra

Create a table with five columns and ten rows as shown in table 1.6.

Table 1.6 An Incomplete Illustration for LCM and GCF

Item No.	Terms	Different Way of Writing Terms from Column # 2	LCM	GCF
1	2, 3	1*2, 1*3		
2	2, 3, 4	1*2, 1*3, 1*4		
3	6, 8	2*3, 2*4		
4	5, 15, 20	5*1, 5*3, 5*4		
5	$1, x, x^2, x^4$	$1*1, 1*x, 1*x^2, 1*x^4$		
6	x^2, x^5, x^8	$1*x^2, 1*x^2*x^3, 1*x^2*x^6$		
7	$2x^3, 6x^5, 4x^4$	$2*x^3, 2*3*x^3*x^2, 2*2x^2*x^2$		
8	$x, (x-1), (x-2)$	$1*x, 1*(x-1), 1*(x-2)$		
9	$2x, 6x(x-2), 8x^2(x-2)$	$2*x, 2*3*x(x-2), 2*4*x*x(x-2)$		

With students' input, complete columns 4 (LCM) and 5 (GCF) in table 6.1. The answers are shown in table 1.7.

Table 1.7 A Complete Illustration for LCM and GCF

Item #	Terms	Different Way of Writing Terms from Column # 1	LCM	GCF
1	2, 3	1*2, 1*3	6	1
2	2, 3, 4	1*2, 1*3, 1*4	12	1
3	6, 8	2*3, 2*4	24	2
4	5, 15, 20	5*1, 5*3, 5*4	60	5
5	$1, x, x^2, x^4$	$1*1, 1*x, 1*x^2, 1*x^4$	x^4	1
6	x^2, x^5, x^8	$1*x^2, 1*x^2*x^3, 1*x^2*x^6$	x^8	x^2
7	$2x^3, 6x^5, 4x^4$	$2*x^3, 2*3*x^3*x^2, 2*2x^2*x^2$	$12x^5$	$2x^3$
8	$x, (x-1), (x-2)$	$1*x, 1*(x-1), 1*(x-2)$	$x(x-1)(x-2)$	1
9	$2x, 6x(x-2), 8x^2(x-2)$	$2*x, 2*3*x(x-2), 2*4*x*x(x-2)$	$24x^2(x-2)$	$2x$

Based on tables 1.6 and 1.7, have students draw conclusions. Some conclusions that can be drawn are as follows:

1. When whole numbers are involved, the least common multiple (LCM) is usually the *highest* number in the series of terms or a *multiple of some of the highest terms*.
2. The greatest common factor (GCF), on the other hand, is the *least number* that is common in all of the terms.

Parallel Organization and Teaching

Parallel organization and teaching is an application of parallel processing from computer science to classroom teaching and learning. As such, it is a new arrow for the pedagogical quiver, and it makes use of analogical reasoning as the link. The idea behind parallel organizing and teaching is that *seemingly individual or different topics or concepts that are often taught separately from one another actually do have some basic feature(s) in common* and that they can be combined and should be presented together.

A good way to organize these kinds of topics and concepts is in tabular form, which immediately suggests they have something basic in common. One such example has been presented earlier in this chapter, in which currency, fraction notation, percent notation, and decimal notation are organized and presented together. Similar concepts need to be combined and presented together for efficiency in teaching and improved student learning.

Another example taken from algebra is factoring and using quadratic equations. Different ways of factoring and the conditions under which the quadratic equation is used need to be organized and presented together. For instance, the expressions shown in table 1.8 should be factored completely.

As shown in the table, the teacher should write information on the board under five columns and then solve each problem in its column. The teacher, then, should discuss the similarities and differences for these five problems. For example, the factors of 10 are 2 and 5, but they can be added and multiplied in different ways (with positive and negative signs) to get the correct answers. Notice, how both integers 5 and 2 are positive during factoring for the first problem, and then they change to -2 and -5 for the problem in the fourth column. The solution for the fifth problem is:

$x = \dfrac{-(-4) \pm \sqrt{(-4)^2 - 4(1)(-10)}}{2} = \dfrac{4 \pm \sqrt{56}}{2} = 2 \pm \sqrt{14}$. This equation has two solutions.

Table 1.8   Teaching Factoring in a Parallel Way

Problem # 1	Problem # 2	Problem # 3	Problem # 4	Problem # 5
$x^2 + 7x + 10$	$x^2 + 3x - 10$	$x^2 - 3x - 10$	$x^2 - 7x + 10$	$x^2 - 4x - 10$
$x^2 + 5x + 2x + 10$	$x^2 + 5x - 2x - 10$	$x^2 - 5x + 2x - 10$	$x^2 - 5x - 2x + 10$	Cannot be factored. Use the quadratic solution to solve for x as discussed in the text for this table.
$x(x + 5) + 2(x + 5)$	$x(x + 5) - 2(x + 5)$	$x(x - 5) + 2(x - 5)$	$x(x - 5) - 2(x - 5)$	
$(x + 5)(x + 2)$	$(x + 5)(x - 2)$	$(x - 5)(x + 2)$	$(x - 5)(x - 2)$	

THE GATEWAY ACADEMIC PROGRAM ASSESSMENT STRATEGY

Four skills or tactics currently make up the assessment strategy, involving the appropriate use of:

1. "Good" errors
2. Exit questions
3. Priming homework
4. Summative tests, quizzes, and examinations

The Good Error and Its Illustration in Basic Math and Algebra

In an earlier publication, we defined typical "good errors" as "partial answers to a question, an answer to something previously learned but tangential to the present class presentation or discussion, and a response or comment that anticipates a future topic or content" (Daryao S. Khatri and Anne O. Hughes, *Color-Blind Teaching: Excellence for Diverse Classrooms* [Lanham, MD: ScarecrowEducation, 2005], 60). In addition, we have found that "good errors" also show that the students are paying attention, are interested in the subject matter, and are not afraid to ask questions, make comments, or offer an answer of some sort that might just hit the mark.

It's up to the teacher to find a way to capitalize on the error, do something productive with it, and thereby enhance the students' grasp of the subject matter. For example, if you ask a class for the cube root of 64, some student probably will provide an answer of 8. This answer is a good error. Remind that student that the *square root* of 64 is 8, but 8 is not the cube root of 64, which has to be a smaller number. This hint will provide students with an opportunity to correct the initial answer, and someone probably will give the correct answer of 4 for the cube root of 64.

Turning good errors into teaching points is not necessarily a skill that can be practiced every day in the Assessment Strategy, because generally it depends on a student's response to initiate its use. However, we have presented it first because its potential benefits to learning are often overlooked by teachers.

Exit Question(s)

Assessment should be more than a summative quiz or test to measure students' learning. Rather, assessment should be integrated into the ongoing instruction to provide regular feedback that can be used to guide further instruction as well as to enhance student understanding. It should be used in multiple formats to measure not only procedural but deeper conceptual understanding.

These formative evaluation questions are presented and answered in a short period of time at a natural breaking point in instruction, and provide immediate feedback about student understanding of the "big picture." They also provide an opportunity to modify instruction and address any widespread confusion or difficulty (George Ashline, "Integrating Exit Questions into Instruction," *News Bulletin*, October 2005, http://www.nctm.org/news/content.aspx?id=618).

We expand a little further on these concepts. Rather than limiting "exit" questions simply to assessing the students' learning, we suggest that they need to be used as a tool to measure the effectiveness of teaching as well. An exit question at an appropriate breaking point in a class session can be either a question or a set of quick questions that tests the conceptual understanding of a certain topic and/or tests the ability of students to apply their new skills for a new concept to a different situation.

The amount of time given to such questions should be limited. That is, they should not take the students more than a few minutes to answer. For example, if you have just finished the concepts of exponents and have taught the basic mathematical operations with simple fractions during previous sessions, the exit questions could be:

$$\text{Simplify: } x^{\frac{1}{2}} \cdot x^{\frac{3}{4}} \text{ and } \left[x^{\frac{2}{3}} \right]^{\frac{3}{2}}.$$

These exit questions will test the understanding of several basic and conceptual concepts, including addition and multiplication of fractions and various exponential rules. If the students answer correctly, you can move on; if not, then you have learned that some reteaching is necessary.

Priming Homework

A *priming homework* problem is usually a simple assignment that alerts the students to the concepts and skills they will need for the *next* course topic, but for which they do possess the necessary information. As the teacher, you can check this homework, and then decide if the concepts and skills required by this topic need review or even reteaching. Without understanding and using the concepts and skills required by a particular topic, students will keep on making mistakes in problem solving.

Students must develop the confidence to problem-solve as well as acquire the mastery of these skills and concepts to be successful, and teachers can provide manageable challenges in this regard. Additionally, the priming homework problem can prompt you as the teacher to quiz students continuously about these skills and concepts, and the students are forewarned about the concepts and skills they must try to have answers for before the next class period.

The overall effect of this approach is that the students now are "primed" or alerted to the need to deal with related new information, and you can present the course content in more effective and quicker ways without creating any frustration on the part of the students. Unlike the straight practice homework that typically appears at the end of each chapter in standard textbooks, the priming homework has to be created for each chapter or new topic by either the teachers themselves or the developers of the course material.

For example, if a teacher wants to discuss the slope of a line and the techniques for obtaining the equation of a straight line, a sample priming homework problem could take the following form:

Simplify.

1. $11 - 7$
2. $7 - 11$
3. $-11 - 7$

Evaluate when $x = 5$, $y = -6$, and $z = 3$.

4. $x + 2(x - y)$
5. $\dfrac{z + x}{y + x}$

6. $\dfrac{x - y}{y + x}$
7. $\dfrac{y - x}{x - y}$

Summative Tests, Quizzes, and Examinations

Everybody is familiar with the time-honored testing tools in the academic repertory. So, what can we add? Well, a few things, we think.

First, always inform the students what is going to be covered on the test. Cover only the most important concepts and skills, that is, those needed for success in the current course or program and those needed as foundation for the next course in the math sequence or requirement. One tactic we have used is to give the students a pool of problems, and tell them we will be choosing a sample from this pool for the test. That way, the students can focus their efforts and practice solving the problems as much as they feel is necessary.

Second, we give bonus points for "special" problems, such as a complex problem for which they must—and actually do—possess all the necessary concepts and skills. Earning the bonus points sometimes helps the students out if they have made a careless error on the main part of the quiz or test.

Third, if there is a departmental examination, we let the students know what its scope and topics are so they can prepare. And we often review material if some of the tasks have not been practiced in a while.

The students' performances on the quizzes and tests are then analyzed with them and provide the bases for reteaching, review, and additional practice.

STUDENT OVERLEARNING

When a student has such a complete mastery of a set of information, concepts, or skills that when that student has need of them, they are instantly available; they are, in effect, *"overlearned"* or *automatic*. The example of the U.S. currency system as the anchoring point for reviewing or introducing new systems is an instance of overlearning.

Success in learning mathematics depends heavily on overlearning or totally mastering the rules (e.g., inverting division problems in fractions, set steps in factoring) and basic systems (e.g., the multiplication tables) for solving problems accurately and efficiently. Just understanding a process or a concept and performing the operations for it a time or two is not synonymous with attaining a total mastery of it.

In making certain our students had complete mastery of the multiplication tables through 15, we lost count of the number of times we went over them, in different ways. We practiced them daily for four or five weeks until all of the students had attained instant recall of any combination.

This strategy is neither as exciting nor as stimulating as the skills of the inductive strategy are. In essence, it requires the teacher to bang away at the needed information until the students can speedily apply it anytime and anywhere. Basically, two skills are involved: *review* and *practice in the classroom*.

Outside-of-class homework does provide some practice, of course, but it does not require the student to perform quickly and independently the way in-class practice, reviews, and tests do. After all, any student can call up another student and confer about the problems assigned, and we encourage our students to do so.

The challenge to the teacher is to find different variations on the theme in reviewing and practicing the necessary information during class time. Here are some variations we have used:

- oral, quick student responses to teacher questions at the beginning of the class period, then with the students later on doing the questions
- handouts that require the students to fill in the information needed for a sampling of the system or concept's applications

- after-class sessions for students who are still struggling a bit
- assignments that require the application of the basic system or concept(s) involved.

Also, we clearly explain—and repeat and repeat and repeat—why the learning of this information is so critical to their success in mathematics and mathematics-based disciplines, such as physics and engineering.

STREAMLINING THE COURSE CONTENT OR PROGRAM

Although this strategy is the last one to be presented in the chapter, you actually should engage in it before the course or the special program begins. The strategy of streamlining is particularly important when, like us, you are trying to close the gap in mathematics in order to ready your students for college-level math courses. The skill cluster for this strategy involves the use of "parallel" curriculum design, described earlier in this chapter, and "pruning."

The implementation of this strategy for math requires the following steps, all of which we carried out:

1. We took the university's particular placement test ourselves. For us, it was the Educational Testing Service's *Accuplacer*, a test widely used at the postsecondary entry level. We wanted to see exactly what types of problems were involved. This test samples basic math, a few statistical applications, and introductory algebra. If your institution uses a different placement test than the *Accuplacer*, we recommend that you take that particular test and analyze the kinds of topics and problems that are sampled. However, it is likely they are similar to those of the *Accuplacer*.
2. We examined departmental syllabi, final examinations, and textbooks for the content being taught in the remedial courses in our university. Regarding these materials, we found the examinations to be most useful.
3. We also took a careful look at the first math courses being offered for college-level credit. The students had to pass Basic Math and Introductory Algebra with acceptable scores in order to be eligible to take the college courses.

Then we began the process of planning the intensive "catch-up" program to close the gap between where students typically were and where they needed to be so they would be able to enroll in a math course at the college level—and succeed in it. This planning process resulted in our eliminating all of the frills that have been included in typical remedial math textbooks, such as puerile and only quasi-relevant pictures, unnecessary content and historical information, problems featuring lots of decimal points to make their solutions extremely tedious, the need for calculators, and excessive numbers of problems and exercises for practice.

We offered our intensive program for the first time during summer 2006 as a research study. Since that time, we have tested, revised, and refined the content and teaching strategies in three additional research studies with groups of diverse students who scored at the bottom of the *Accuplacer* at the beginning of the program. At the end of the program, the majority of these students became eligible for college math courses upon their second testing with the *Accuplacer*.

We followed up those who were eligible for college-level math courses and enrolled in them. This group of students was able to succeed in college-level math upon the completion of our program.

What is presented in this book then is the intensive math program that we have researched and refined over a period of four years. However, we hasten to point out that the teaching strategies with their skill clusters that are presented in this chapter using math as the content have been studied, analyzed, tested, and refined over a period of many, many years for previous summer programs and for other regular courses we have taught as well.

POINTS TO REMEMBER

- The management rules must be enforced. Without all of the students' undivided attention, your teaching effectiveness will be seriously diminished. With students who are not used to paying attention, you may have to spend a good bit of time going over and enforcing the rules. Explain that class time is precious, and this experience is not a sequel to television's law-and-order programs.
- Learn the students' names as fast as possible. Knowing their names is key to class control and the inductive strategy's skills of differential knowledge bases and equal participation.
- Exit questions, assigning and checking practice and priming homework are critical components for students' mastery of the subject matter and their ultimate success.
- Preparing a daily handout containing teacher-directed work, students' independent class work, exit questions, and priming homework is important. This handout establishes a routine; it keeps everyone on track. Also, it will help to pinpoint where reteaching and/or review of a particular topic will be needed.
- Even if you as the teacher feel a little uncomfortable at the beginning with the management rules and the skills of the inductive strategy, keep us-

ing the ones you feel are the most important. The students will also very quickly register that you are doing your level best to teach them a subject they need to know to succeed in college.

In the next chapter, we show a sequence of several class presentations of the pruned math content that involve the five teaching strategies and their skill clusters.

2
Finding Gaps and Honing Basic Skills

In this chapter, we provide a suggested set of sequentially arranged math activities for the first several days of the intensive short-term math project called the Gateway Academic Program (GAP). This program is designed to assist the teachers in several ways. First, the sequencing of the problems allows the teacher to establish quickly the floor of the students' knowledge base and to find the gaps in students' knowledge bases in basic math and algebra at the very beginning of the course. Second, students' sharpness of responses in terms of basic skills to the problems provides quick feedback to the teachers as to how fast and how far the students can be pushed to achieve all the course goals. Third, the basic math and algebra problems are sequenced, practiced, and reviewed in an inductive way to assure students' successes every day. In actuality, the teacher will find that the project picks up speed as the students continue to experience success.

Specifically, the sections of this chapter begin with an overview of a preprogram orientation that a teacher can use for both the parents and participating students to maximize both student and parental understanding of the program. Next, the intensive teaching model that has been found effective with these young adults is briefly described as well as the time blocks involved.

The main body of the chapter is given over to presenting:

1. those teaching strategies and skill clusters of chapter 1 that a teacher should find helpful with the all of the math content,
2. a set of problems sequenced in an inductive way that a teacher can use in direct instruction (teacher-directed instruction),

3. a complementary set of problems for students to solve (student-directed class work), and
4. a proposed organization of a typical four-hour day in six different time blocks.

PRE-PROGRAM ORIENTATION FOR COMPRESSED PROGRAMS

The pre-program orientation should be organized so that everyone will be clear about the objectives of the program, the hours involved, the student incentives, and the anticipated outcomes. Parents are interested in what is going to take place, and they often want to know what, if any, direct role they may have in the process. Also, they find it reassuring to meet the teaching faculty and student assistants and to see the classroom and other facilities involved.

Serving refreshments and introducing other educational officials also serves to give credence to the program as well as providing an opportunity for informal interchanges. If at all possible, some monetary or scholarship incentive should be offered to the participating students to offset not being able to work during the months of the summer program.

For a program occurring in the regular semester or quarter pattern, the orientation is not so important unless there are marked variations in the number of hours involved or in the program goals and requirements.

Toward the end of the orientation, teachers are encouraged to hold an individual session with each student to explain that student's placement test results. This session is particularly important if a student's high school grades in math were all passing. Students need to have an understanding of what their particular scores and percentile ranks mean in terms of college performance. Prior to the orientation, all of the participating students in the program must have taken whatever the institution's approved placement test is, such as the College Board's *Accuplacer*.

ORGANIZATION OF INSTRUCTION AND TESTING IN A TRAINING MODEL

The term *faculty trainer* or *trainer* is used here because it carries a different connotation than the term *faculty* or *teacher*. The term *trainer* derives from typically short-duration, intensive, contract-funded projects with very specific, easily measured, and limited sets of "deliverables" or objectives.

In short, the trainer seeks to provide the adult trainees with a particular skill or set of skills at a clearly defined level of measurable proficiency, such as advanced training in the use of a spreadsheet or a word-processing program. For the Gateway Academic Program in math, we adapted this training model to a pruned-to-essentials mathematics catch-up program for use with incoming, seriously underprepared entering college freshman students.

The training model has several advantages:

1. It is designed to be used with adults, and the students are treated as young adults—not as unruly high school students.
2. The objectives or "deliverables" are very specific (i.e., to pass both the university's placement test and the university's departmental examinations in math at the critical levels for admission to a college-level math course).
3. A daily agenda that specifies the activities for the day is used.
4. Handouts are used as instructional material rather than textbooks with chapters.
5. The classroom management rules (noted in the previous chapter) are enforced on the grounds that high school is history, and they are adults now.

Program Time Frame and Daily Organization

In a training program, the organization is critical and should follow a set pattern. The program model presented here has been designed for eight weeks, four hours per day, Monday through Friday, for a total of 160 hours. A suggested daily organization of four hours for teachers follows.

Each teaching day Monday through Thursday is divided into six time blocks.

Time Block #1	10:00–11:15 AM	Teacher-Directed Instruction I
Time Block #2	11:15–11:45 AM	Student Independent Class Work I
Time Block #3	11:45 AM–Noon	Break I
Time Block #4	Noon–1:15 PM	Teacher-Directed Instruction II
Time Block #5	1:15–1:45 PM	Student Independent Class Work II
Time Block #6	1:45–2:00 PM	Short individual consultations

All days of the week, except Fridays, should follow the same schedule of six time blocks. The Friday time block is devoted to a review of the week's work, followed by a comprehensive test. For each day, under the heading "Teacher-Directed Instruction," the problem sets or topic listings noted for that day also can serve as agendas. In other instances, only the topics to be covered in the teacher-directed instruction are listed, and additional handouts or board work should be used to provide examples and practice. Note taking by students, including all of the completed student handouts, should be placed in students' three-ring binder notebooks and should be checked by the teacher or a student assistant on a regular basis as a part of developing needed study habits and organization.

Within these time blocks, the following activities should be carried out at appropriate times during the day.

1. Check practice and priming homework.
2. Take attendance.
3. Pass out the agenda for the day.
4. Announce which students did not complete their homework. They must complete it on that day. (This rule must be enforced for 100% compliance. Otherwise, the students will slack off.)
5. Answer questions about the homework.
6. Review previous work and "must know" topics for the course.
7. Use the inductive strategy to introduce new topics, and the skills of the differential knowledge bases of students to assure equal participation for all students.
8. Use exit questions to assess the effectiveness of the teaching for that day.
9. Assign practice and priming homework and/or class work.
10. Confer frequently with other colleagues who are participating in the program at the end of the completed sessions for the day in a "How are we doing?" meeting. These meetings should follow up on students' responses to the exit questions and allow for planning any needed adjustments in the topics for the subsequent days.

Class Management Rules

In the training model, the "rules of the road" for teacher–student interactions are laid down at the beginning of the program. The importance of these rules to the students' successful completion of the program must be stressed—and enforced—and the students should agree to follow them. Have the students read these rules aloud on a rotational basis. In the early stages of the program, the teacher should absolutely enforce whatever the rules are—no matter how often this enforcement must take place. Without such enforcement, much valuable teaching time will be lost, and the entire program will be in jeopardy.

1. Be in the class *on time*; don't be late.
2. Calling in about being sick is not an option; rain or shine, you show up.
3. Treat this program as work; no pay for days you miss.
4. Deposit cell phones and iPods at the door before you enter the classroom.
5. Use restrooms *before* you come to the class.

1. *Dealing with positive and negative integers.* These problems are designed to create three basic skills in students. First, if both integers are positive, add the two integers and insert a *positive sign* in the answer. Second, if both integers are negative, add the two integers and insert a *negative sign* in front of the answer. Third, if one integer is positive and the other negative, subtract the lower integer from the higher one, and *insert the sign of the larger integer* for the answer. The teachers are encouraged to test these skills every day until all students can demonstrate them.
   1.1. Addition of integers ($8 + 4$, $11 + 19$, and $-5 + 3$)
   1.2. Subtraction of integers ($11 - 7$ and $8 - 13$)
   1.3. More on subtraction of integers ($-5 - 6$ and $-9 - 11$)

2. *Reviewing the properties of 0 and 1.* The teacher should encourage all students to overlearn properties of zero and one. First, if you add or subtract a zero from an answer, it does not change the answer. Second, if we multiply with zero, the answer is zero. Third, *you cannot divide by zero*. Fourth, division or multiplication by one does not change the answer.
   2.1. Addition and subtraction of 0 ($11 - 0$ and $22 + 0$)
   2.2. Multiplying and dividing by 0 ($\frac{0}{5}, \frac{0}{256}, \frac{5}{0}, \frac{1234}{0}$)
   2.3. Multiplying and dividing by 1 (12×1 and $567 \div 1$)

3. *Squares and square roots.* The teacher should make certain that students overlearn the squares and square roots of some common integers.
   3.1. Ask for the squares of simple integers (4, 5, 6, 7, 8, 9, and 10).
   3.2. Ask for the square roots of 4, 9, 16, 25, 36, 49, 64, 81, and 100.

4. LCM. Alert students that, although LCM means *least* common multiple, the answer for an LCM is usually the *largest* number or term in the series or a multiple of some of the highest numbers or terms. The LCM is always divisible by all terms in the sequence.
   4.1. Find the LCM of 4 and 12.
   4.2. Find the LCM of 4, 6, and 8.

5. *Writing large and small numbers in powers of 10.* Every student should learn that an equal number of trailing zeros in the numerator and denominator terms can be canceled when performing division of integers.
   5.1. 1,000,000; 100,000,000; 1,000,000,000,000,000
   5.2. 0.000001 and 0.000000000001
   5.3. Divide by 10, 100, 1000 (230/10; 4500/100, etc.)
   5.4. Write 657.89 in scientific notation

6. *Dealing with simple fractions.* The skill to be learned here is that when dividing a fraction with an integer, change the division *line* to a division *sign*. Then change the division sign to a multiplication sign, replace the integer or fraction that follows the division sign with its reciprocal, and then simplify.

   6.1. Adding ($\frac{1}{2} + \frac{1}{2}$; $\frac{1}{4} + \frac{3}{4}$)

   6.2. Subtraction ($1 - \frac{1}{2}$; $\frac{1}{2} - \frac{1}{4}$)

   6.3. Dividing ($\frac{\frac{1}{2}}{2}$; $\frac{4}{\frac{1}{2}}$)

   6.4. Multiplication ($\frac{3}{5} \times \frac{10}{3}$ and $\frac{5}{25} \times \frac{50}{10}$)

Student Independent Class Work I (Day 1)

Half an hour should be allowed for the students' independent class work, which is shown below. While the students should finish the class work independently, they must receive help from the teacher if some of them become stymied and need assistance. They should not ask one another for help; they need to ask a teacher.

The idea here is that every student should succeed right from the start. The student independent class period is not to be considered as a test, but rather as practice. Also, the teacher should walk around and observe the students' work habits as they work on the problems.

After finishing their class work, students should take a break for 15 minutes.

Add, subtract, multiply, and divide.
1. 5 + 11
2. 11 + 19
3. −19 + 9
4. 21 − 11
5. 13 − 16
6. 4 × 9
7. 4 × (−10)
8. (−7) × 8
9. (−7) × (−9)
10. (−4) × (−11)
11. 11 × 11
12. 13 × 13
13. 9 × 9
14. $\dfrac{22}{11}$
15. $\dfrac{-33}{3}$
16. $\dfrac{15}{-5}$
17. $\dfrac{-50}{-25}$

Properties of 0
18. Add: (103 + 0)
19. Subtract: (2 56 − 0)
20. Multiply: 1024 × 0; y × 0; (100 − 3z) × 0
21. Divide: $\dfrac{51}{0}$; $\dfrac{256}{9-9}$; $\dfrac{2048}{y-y}$
22. Simplify: 10^0; x^0; $(y - 100)^0$

Properties of 1
23. Multiply: 9 × 1 ; 345 × 1; (10z − y) × 1 ; {1 × 267^0 × $(2y)^0$}
24. Divide: $\dfrac{11}{1}$; $\dfrac{689}{1}$; $\dfrac{9y - 3p}{1}$
25. Simplify: (10^1___; x^1___; $(5z)^1$___)

Divide by 10, 100, 1000, etc.
26. $\dfrac{50}{10}$
27. $\dfrac{500}{100}$
28. $\dfrac{6000}{1000}$
29. $\dfrac{50,500}{100}$
30. $\dfrac{0,050,500}{10,000}$
31. $\dfrac{105,000}{1000}$

Teacher-Directed Instruction II (Day 1)

In addition to providing assistance as needed during the time the students are completing the independent block of work, the teacher also should answer any additional questions that the students have regarding problems they may have encountered in completing their work in this time block. The teacher should again use the equal participation teaching skill in working through the topics in this section to further identify gaps in the students' knowledge of these topics. The teacher should use the printed material under this section as an agenda.

1. *Checking on the measurement of angles of triangles.* The teacher should describe the importance of two angle measurements, 90° and 180°, when dealing with triangles.
   1.1. The sum of the measurements of the angles of a triangle
   1.2. The measurement of three angles of an equilateral triangle
   1.3. Isosceles triangles—one angle is given, how to find the other two angles
   1.4. Right triangles
2. *Intersection of lines.* The teacher draws intersecting lines before asking the question. *Vertical angles* are opposite to each other; *adjacent angles* are next to each other; any two angles whose sum is 90° are *complementary*; any two angles whose sum is 180° are *supplementary*. These definitions escape many students over a couple of days. The teacher should make a habit to review and quiz students on these concepts periodically; it takes no longer than 30 seconds to review this information.
   2.1. Vertical angles
   2.2. Adjacent angles
   2.3. Complementary angles
   2.4. Supplementary angles
3. *The Rectangular Coordinate System.* The teacher needs to emphasize that angles are measured counterclockwise, and that the quadrants are labeled counterclockwise as well. Students need to make the connection with the quadrants and the angle measurements of 90°, 180°, 270°, and 360°. Finally, the teacher needs to review the x- and y-values in terms of their signs in each quadrant.
   3.1. The x- and y-coordinates of a point
   3.2. Locating a point on the coordinate system if its values are given

Student Independent Class Work II (Day 1)

Again, approximately half an hour should be allowed for the class work. The students should finish the class work independently if they can. However, if someone is having trouble with a problem, then that student should receive help from a teacher as needed. After finishing their class work, students should have their work immediately checked, so they will know how successful they have been.

Addition, subtraction, multiplication, and division involving integers.

1. $5 + 11 + 4$
2. $11 + 5 + 4$
3. $-19 + 9 + 10$
4. $21 - 11 - 11$
5. $13 - 16 + 3$
6. $4 \times 9 \times (-1)$
7. $4 \times (-10) \times (-2)$
8. $(-7) \times 8 \times (-1)$
9. $(-8) \times (-9)$
10. $(-6) \times (-11)$
11. $11 \times 11 \times (-1)$
12. $13 \times 13 \times (-1)$
13. $12 \times 9 \times (-1)$
14. $\dfrac{-22}{11}$
15. $\dfrac{-33}{-3}$
16. $\dfrac{20}{-4}$
17. $\dfrac{-75}{-25}$

Properties of 0
18. Add: $(103 \times 0 + 5)$
19. Subtract: $(256 \times 0 - 1)$
20. Multiply: 11×0; $z \times 0$; $(1200 - 36z) \times 0$
21. Divide: $\dfrac{25 * 12 * 11}{9 - 9}$; $\dfrac{5 * 2 * 6}{9y - 9y}$
22. Simplify: 100^0; $(z - 5)^0$

Properties of 1
23. Multiply: 11×1; 45×1; $(10z - 9z) \times 1$; $\{3 \times 267^0 \times (2y)^0\}$
24. Divide: $\dfrac{21}{1}$; $\dfrac{689}{1}$; $\dfrac{9p - 8p}{1}$
25. Simplify: 100^1 ___; z^1 ___; $(11z)^1$ ___

Dividing by 10, 100, 1000, etc.

26. $\dfrac{50}{10}$
27. $\dfrac{100}{10 * 10}$
28. $\dfrac{6000}{100 * 10}$
29. $\dfrac{50,500}{10 * 10}$
30. $\dfrac{00,050,000}{100 * 100}$
31. $\dfrac{205,000}{2 * 10 * 50}$

Every student must answer every problem in the sets correctly or the teacher must consider that a gap does exist—at least for some student(s). A majority is insufficient.

Day 2: Identifying Students' Knowledge Gaps, Continued

The major activity for day 2 of the program continues to be the identification of gaps in students' knowledge bases in basic math and introductory algebra. The teacher should continue to use the students' differential knowledge bases and equal participation skills for determining these gaps.

Teacher-Directed Instruction I (Day 2)

As before, the teacher should ask students a question or encourage them to comment on a question or concept on a random basis when discussing this section.

1. *Equation of a straight line.* The teacher should constantly reinforce the standard equation of a straight line and apply it in situations where the coefficient of y is not 1. Try to use the negative sign liberally for different terms to provide practice to students in writing these equations in standard form before identifying the slope and the y-intercept.
   1.1. Slope of a line; draw a line first that has a y-intercept
   1.2. Standard equation of a straight line ($y = mx + b$)
   1.3. Identify the y-intercept and the slope of the line in: $2y = -6x + 7$

2. *Solve for x in simple equations.* The simple principle here is to collect like terms and, when transferring terms across the equal sign, change the sign of that term. The teacher should *stay away* from using the concepts of adding and subtracting terms from each side to solve such equations. Students have a tendency to make additional mistakes with this approach.
   2.1. $3x = 27$
   2.2. $5x + 3 = 11 + x$

3. *Solve for x using cross products/cross multiplication.* It is very important that a teacher emphasize simplification before cross multiplying terms. Students have problems with this technique because they usually end up with large coefficients if they do not simplify. As soon as they see large coefficients, they get discouraged and very often quit.
   3.1. $\dfrac{2x}{9} = \dfrac{4}{3}$
   3.2. $\dfrac{2x}{20} = \dfrac{5}{2x}$

4. *Simplify.* Students should have mastered—overlearned—the square roots of some standard numbers before they attempt to tackle this set of problems. If they have not, review the square roots of some standard positive integers. Let students know that the square root of a power is simply half of that power—which amounts to multiplying the power by ½. Again make certain that students simplify terms if both the numerator and denominator are present before taking the square root of terms inside the radical.

4.1. $\sqrt{x^{10}}$

4.2. $\sqrt{36x^4}$

4.3. $\sqrt{\dfrac{4900x^5y^9}{100xy^3}}$

Student Independent Class Work I (Day 2)

Addition of negative terms

1. $-8 - 6 =$
2. $-11 - 6 =$
3. $-9 - 2 =$
4. $-14 - 12 =$
5. $-8 - 2 =$

Addition and subtraction with multiple terms

6. $6 - 2 - 2 =$
7. $12 - 4 - 6 =$
8. $9 - 2 - 5 - 2 =$
9. $8 - 2 - 1 - 1 =$
10. $9 - 1 - 7 - 2 =$

Division with single terms

11. $25/5 =$
12. $36/6 =$

13. $22/11 =$

14. $24/3 =$

15. $45/9 =$

Addition and subtraction involving parentheses

16. $-12 + (17 - 5) =$

17. $-15 + (22 - 7) =$

18. $-8 - (15 - 7) =$

19. $-10 - (22 - 20) =$

20. $-5 + (17 - 12) + 10 =$

Simplify:

$$\frac{55 \times 36}{11 \times 6} =$$

$$\frac{24 \times 36}{6 \times 8} =$$

$$\frac{28 \times 64}{7 \times 8} =$$

Teacher-Directed Instruction II (Day 2)

The teacher continues to assess students' knowledge bases and gaps in the areas of polynomials and factoring using the equal participation teaching technique. The problem set also serves as an agenda for the session.

1. *Addition and subtraction of polynomials.* Students must *rewrite* the problem with similar terms written next to each other (together) before adding or subtracting the coefficients of these terms. Powers of the variables are not impacted in this process.
   1.1. $5x + 3x + 11 + 2x - 1$
   1.2. $5x^3 - 2x^2 + 3x^3 + 3x^2 + 6x$

2. *Factoring.* First, students need to factor something that is common in all of the terms before carrying out the final act of standard factoring. The common factors are *usually* the lowest coefficients and terms with the lowest powers.
   2.1. Factor: $4x^3 + 12x^4$

2.2. Factor: $x^2 - 1$
2.3. Factor: $25 - 16x^2$

3. *Multiplying polynomials.* In multiplying polynomials, students need to focus on three aspects, in a set order. First, take care of the signs of the terms. Second, multiply the coefficients. Third, multiply the terms with powers. The order is important because students have a tendency to overlook the signs in multiplication.
   3.1. Multiply: $3(3x^2 + 4x + 6)$
   3.2. Multiply: $(x - 3) * (-2x + 4)$

4. *Simplify.* Here, factor first, then cancel like terms in the numerator and denominator next. The very important rule to remember is that the denominator as written in 4.2 acts as a denominator for *all* of the terms. The common mistake is that many students and professionals have a tendency to cancel the $5x$ in the numerator with the $5x$ in the denominator and leave the rest of the numerator intact, which, of course, is wrong.
   4.1. $\dfrac{(x^2 - 4)}{x + 2}$
   4.2. $\dfrac{(5x^5 + 10x^3 + 25x^2 - 5x)}{5x}$

Student Independent Class Work II (Day 2)

The problem set printed below is used as an independent class work for students after teacher-directed instruction II.

Addition of negative terms
1. $-12 - 5 =$
2. $-11 - 4 =$
3. $-9 - 1 =$
4. $-8 - 2 =$
5. $-10 - 3 =$

Addition and subtraction with multiple terms
6. $-2 - 3 + 7 =$
7. $-5 - 4 + 8 =$

8. $-3 + 3 - 5 + 5 =$

9. $-6 + 2 + 5 - 2 =$

10. $-7 + 2 - 2 + 7 =$

Division with single terms

11. $56/8 =$

12. $99/11 =$

13. $36/6 =$

14. $72/8 =$

15. $64/8 =$

Addition and subtraction involving parentheses

16. $(-12 + 17) - 5 =$

17. $-15 + (20 - 5) =$

18. $-11 - (15 - 4) =$

19. $-10 - (22 - 12) =$

20. $-5 - (17 - 12) - 10 =$

Simplify:

$$\frac{18 \times 24}{4 \times 6} =$$

$$\frac{45 \times 16}{9 \times 8} =$$

$$\frac{44 \times 50}{11 \times 10} =$$

Day 3: Review of Basic Math Departmental Final Examination

The teacher should review the university's final exam (the exam that is administered to all students taking remedial basic math) using the inductive approach to test the students' proficiency in the basic skills of arithmetic. The results of this informal testing will allow teachers to determine how quickly the gaps in basic math can be closed and how far and how fast they can push these students in learning these and new concepts of this course.

Finding Gaps and Honing Basic Skills 39

Teacher-Directed Instruction I (Day 3)

The problem set printed here serves as the agenda for the day. The teachers are encouraged to use the skills cluster of the inductive approach strategy, specifically students' differential knowledge bases, anchoring, "good errors," and equal participation to teach all math segments for the day. Comments in the brackets along with a problem are included to signal either the emphasis or the skill that the students must learn.

1. Round 3.5746 to the nearest hundredth [to test rounding off]
2. Add: 312.4 + 8.7 + 93 [to test the decimal alignment]
3. Subtract: 3 − 0.064 [to test decimal alignment and techniques in borrowing]
4. A group of seven friends spent $35.35 on lunch. What was each person's share? [to provide practice on word problems and concepts building]
5. Multiply 66 × 0.126 [to check mastery of multiplication tables and placement of decimals in the final answer]
6. Divide: $\frac{86.7}{0.1}$ [to test moving of decimal to the left or right]
7. Simplify: $2^4 - 30 \div 5$ [to test the understanding of rules for arithmetic operations]
8. What is the condition of divisibility by 6 and 2? [to test rules of divisibility]
9. Simplify: $\frac{48}{40}$ [to test quickness in recognizing the decimal movement and simplification]
10. Multiply and simplify: $\frac{4}{6} \cdot \frac{12}{8}$ [to test for simplification before actually multiplying]
11. Divide and simplify: $30 \div \frac{6}{7}$ [reinforcement of divisibility rules]
12. It takes ⅔ cup of sugar to make one batch of oatmeal cookies. How much sugar is required for 18 batches? [to provide practice in word problems and concepts building]
13. Find the LCM of 9 and 6 [to test mastery of LCM]
14. Subtract and simplify: $\frac{4}{8} - \frac{3}{12}$ [to test the rules in subtraction of fractions and finding LCM]
15. Convert $\frac{47}{8}$ to a mixed numeral [to test longhand division skills]
16. Subtract: $14 - 6\frac{7}{9}$ [to emphasize the subtraction of whole numbers, followed by borrowing, if needed, and then subtraction of a fraction from one. The teacher needs to illustrate all parts of the problem].

17. Multiply: $3\frac{3}{4} \cdot 2\frac{2}{3}$ [changing first to improper fractions from mixed numerals before simplifying and multiplying]
18. Find the average of 6, 9, and 12.
19. Find the average of $\frac{3}{2}, \frac{3}{5},$ and $\frac{3}{10}$ [instruct students not to forget division by three as the last step in solving this problem]
20. Write decimal notation for $11\frac{19}{100}$ [quick recognition of division by 100 as a clue to move decimal places to the left]

Student Independent Class Work I (Day 3)

Students usually have 30 minutes to finish the class work. If they need assistance, they can ask the teacher to help them. This is practice class work and, therefore, the problem set is very similar to that in the section on teacher-directed instruction.

1. Round 8.5796 to the nearest tenth.
2. Add: 212.4 + 8.1 + 93
3. Subtract: 4 − 0.084
4. A group of six friends spent $30.30 on lunch. What was each person's share?
5. Multiply: 63 × 0.14
6. Divide: $\frac{468.9}{0.1}$
7. Simplify: $3^4 - 24 \div 6$
8. What is the condition of divisibility by 6 and 3?
9. Simplify: $\frac{45}{81}$
10. Multiply and simplify: $\frac{4}{5} \cdot \frac{10}{18}$
11. Divide and simplify: $20 \div \frac{5}{6}$
12. It takes ⅔ cup of sugar to make one batch of oatmeal cookies. How much sugar is required for 15 batches?
13. Find the LCM of 6 and 15.
14. Subtract and simplify: $\frac{3}{8} - \frac{3}{12}$
15. Convert $\frac{45}{8}$ to a mixed numeral.

Finding Gaps and Honing Basic Skills

16. Subtract: $15 - 6\frac{8}{9}$

17. Multiply: $3\frac{3}{5} \cdot 3\frac{1}{3}$

18. Find the average of 30, 50, and 70

19. Find the average of $\frac{1}{2}, \frac{2}{5},$ and $\frac{3}{10}$

20. Write the decimal notation for $23\frac{319}{1000}$

Teacher-Directed Instruction II (Day 3)

The teacher should continue with the review of the basic math departmental final examination. The topics and associated problems are listed here. The problem set should serve as an agenda for the class. The comments below are provided to assist the teacher to test students' knowledge or to provide the emphasis of the problem in building skills.

1. Find decimal notation: $\frac{21}{5}$ [to test long division and quickness in simple divisions]
2. A 20-oz. bottle of soap costs $6.00. Find the unit price in cents per ounce. [to provide practice with word problems and concepts building]
3. Find the percent notation for 0.208 [to emphasize moving the decimal to the right]
4. To get an A in Algebra II, a student must average 90 points on four tests. Scores on the first three tests were 86, 89, and 90. What is the lowest score that the student can make on the last test and still get an A? [to provide additional practice in word problems and concepts building]
5. Find the median of the following numbers: 9, 9, 16, 24, 30, 42 [students are usually familiar with the term *median*]
6. Find the mode of the following numbers: 17, 19, 37, 49, 22, 22 [students are usually familiar with the concept of mode]
7. Complete: 6 in = _____ ft [Provide constant practice with conversions]
8. Complete: 2.8 m = _____ cm

The rest of the problems in this section use formulas that a teacher needs to write on the board. Warn students to be careful when using a square or cube of radius in a formula—instruct them to take the square or cube of a number when using such formulas.

9. Find the perimeter of a rectangle that is 12 cm by 7 cm.
10. Find the area of the given polygon (triangle) that has a height of 12 cm and base of 10 cm.

11. Find the area of a given polygon (rhombus) that has a height of 8 ft if the bottom side is 12 ft and the top side is 8 ft.
12. Find the diameter of a circle that has a radius of 11 in.
13. Find the circumference in terms of π of a circle that has a radius of 18 in.
14. Find the area in terms of π of a circle that has a radius of 11 cm.
15. Find one side of a right triangle that has the hypotenuse of 10 cm and whose other side is 6 cm.
16. Find the volume in terms of π of a sphere that has a radius of 3 cm.

Student Independent Class Work II (Day 3)

1. Find decimal notation: $\frac{19}{5}$
2. A 20-oz. bottle of soap costs $5.00. Find the unit price in cents per ounce.
3. Find the percent notation for 0.348.
4. To get an A in Algebra II, a student must average 90 points on four tests. Scores on the first three tests were 85, 90, and 88. What is the lowest score that the student can make on the last test and still get an A?
5. Find the median of the following numbers: 9, 9, 15, 23, 30, 42
6. Find the mode of the following numbers: 17, 19, 37, 49, 20, 20
7. Complete: 9 in = _____ ft
8. Complete: 3.8 m = _____ cm
9. Find the perimeter of a rectangle that is 16 cm by 8 cm.
10. Find the area of the given polygon (triangle) that has a height of 10 cm and base of 20 cm.
11. Find the area of a given polygon (rhombus) that has a height of 9 ft if the bottom side is 13 ft and the top side is 9 ft.
12. Find the diameter of a circle that has a radius of 9 in.
13. Find the circumference in terms of π of a circle that has a radius of 22 in.
14. Find the area in terms of π of a circle that has a radius of 12 cm.
15. Find one side of a right triangle that has the hypotenuse of 15 cm whose other side is 9 cm.
16. Find the volume in terms of π of a sphere that has a radius of 4 cm.

POINTS TO REMEMBER AND REVIEW

- In any mathematical operation, instruct the students always to deal with the sign first, followed by the coefficients, and then the variables.
- In dealing with negative terms and/or negative signs for the same term, make certain the students overlearn the rules as presented here:

  ✓ First, if both integers are positive, add the two integers and insert a *positive sign* in the answer.
  ✓ Second, if both integers are negative, add the two integers and insert a *negative sign* in front of the answer.
  ✓ Third, if one integer is positive and the other negative, *subtract* the lower integer from the higher one, and *insert the sign of the larger integer* for the answer.
  ✓ And then practice, practice, and practice until all of the students' responses are automatic.

- Use the inductive strategy and the skills cluster that goes with it in the teacher-directed instruction.
- Pay special attention to what students actually do know and the topics with which they are having difficulty. Try to know the strengths and weakness of each and every student as soon as possible.

3
Mental Math

When you see students in any grade above the seventh use a calculator to simplify (−5 − 11), to find the square root of 64, or to find the square of 7, you know that these students need our help. In word problems, when students have to ask another student or use a calculator to determine the discount on a dress that is 20 percent off the regular price of $200, something has fallen through the cracks somewhere in math education. Obviously, such students will revert to a calculator to perform simple tasks even in basic math.

And, if you, as a teacher, allow the use of calculators to do addition of even two single-digit numbers, the students' brains are not being exercised, only their calculators and their fingers. However, when students use calculators to perform complex tasks that can save them agonizing amounts of time, then they are a terrific help.

This chapter, then, is aimed at such brain exercise and mastery. It emphasizes the multiplication tables and some simple fractions, decimals, and percentages involving money that are part of everyday use. It also looks at the special rules governing the use of 1 and 0.

MULTIPLICATION TABLES

Just as A, B, C, D, and so on are the building blocks for the words of the English language, mental knowledge of the multiplication tables is the foundation of quick addition, multiplication, subtraction, and division in math. The quick recall of specific combinations from the multiplication

Table 3.1 A Typical Multiplication Table

Whole Numbers (2-15)	2	3	4	5	6	7	8	9	10	11	12	13	14	15
2	**4**	6	8	10	12	14	16	18	20	22	24	26	28	30
3	6	**9**	12	15	18	21	24	27	30	33	36	39	42	45
4	8	12	**16**	20	24	28	32	36	40	44	48	52	56	60
5	10	15	20	**25**	30	35	40	45	50	55	60	65	70	75
6	12	18	24	30	**36**	42	48	54	60	66	72	78	84	90
7	14	21	28	35	42	**49**	56	63	70	77	84	91	98	105
8	16	24	32	40	48	56	**64**	72	80	88	96	104	112	120
9	18	27	36	45	54	63	72	**81**	90	99	108	117	126	135
10	20	30	40	50	60	70	80	90	**100**	110	120	130	140	150
11	22	33	44	55	66	77	88	99	110	**121**	132	143	154	165
12	24	36	48	60	72	84	96	108	120	132	**144**	156	168	180
13	26	39	52	65	78	91	104	117	130	143	156	**169**	182	195
14	28	42	56	70	84	98	112	126	140	154	168	182	**196**	210
15	30	45	60	75	90	105	120	135	150	165	180	195	210	**225**

tables in solving problems provides confidence, quickness, and accuracy for students. Your students need to be tested and quizzed and, where needed, provided written and oral practice on the mental use of multiplication tables every day until they have memorized the 15 × 15 multiplication table grid. A complete 15 × 15 multiplication table is provided in table 3.1, followed by blank homework sheets for practicing at the end of the chapter.

Mental Exercise 3.1

Ask the students to perform the following operations orally.

1. $3 + 7$
2. $9 + 11$
3. $21 + 29$
4. $11 - 4$
5. $21 - 11$
6. $52 - 12$
7. 6^2
8. 9^2
9. 11^2
10. $\sqrt{9}$
11. $\sqrt{36}$
12. $\sqrt{81}$

CURRENCY AND ITS RELATIONSHIP TO DECIMALS, FRACTIONS, AND PERCENTAGES

When students cannot perform simple problems with decimals or fractions like $(0.75 + 0.25)$, $(0.75 - 0.50)$, $(1\frac{1}{4} - \frac{1}{4})$, or $\left(\frac{1}{2} \div 2\right)$ correctly without a calculator, it is time that this information is permanently stored in their brains in a practical way. This section provides a way to do just that.

Every student understands money, and the U.S. monetary system, like many other currencies in the world, is based on fractions or multipliers of 10 and/or 100. The percent notation is calculated by *multiplying* the decimal value by 100, whereas the decimal value can be obtained by *dividing* the value in percent by 100. These computations are shown in table 3.2. Then have students complete table 3.3 (shown over page).

Table 3.2   Dollar Currency and its Fractional, Decimal, and Percent Notations

Money words	Money	Fractional Notation	Decimal Notation	Percent Notation
Dime	10 ¢	$\frac{1}{10}$	0.10	10%
Nickel	5 ¢	$\frac{1}{20}$	0.05	5%
Penny	1 ¢	$\frac{1}{100}$	0.01	1%
Quarter	25 ¢	$\frac{1}{4}$	0.25	25%
Half a dollar	50 ¢	$\frac{1}{2}$	0.50	50%
Three quarters	75 ¢	$\frac{3}{4}$	0.75	75%
Dollar and a quarter	125 ¢	$1\frac{1}{4}$	1.25	125%
Dollar and a half	150 ¢	$1\frac{1}{2}$	1.50	150%
Dollar and 75 cents	175 ¢	$1\frac{3}{4}$	1.75	175%

Table 3.3 An Exercise in Fractional, Decimal, and Percent Notations

Mental Exercise 1.2 Write the fractional, decimal, and percent notations.

Money Words	Money in Cents	Fractional Notation	Decimal Notation	Percent Notation
Two dimes				
10 nickels				
55 pennies				
Two half dollars				
Dollar and a dime				
Dollar and a nickel				

FACTS ABOUT 1 AND 0

Mental Exercise 3.2: Facts about 1 and 0

- Any number divided by 1 is the same number, e.g., $5 \div 1 = 5$
- Any number multiplied by 1 is the same number, e.g., $9 \times 1 = 9$
- Any nonzero number divided by itself is 1, e.g., $7 \div 7 = 1$
- Any number multiplied by 0 is 0, e.g., $21 \times 0 = 0$
- Zero added to or subtracted from any number does not change the number, e.g., $31 + 0 = 31$ and $46 - 0 = 46$
- Zero cannot divide either a number or a quantity. If you try to divide any number or a quantity by zero, it is undefined. In the sciences, we define it as infinity (∞) for simplicity.
- Any quantity other than 0 raised to the power of 0 is 1, e.g., $6^0 = 1$ and also $x^0 = 1$.

Mental Exercise 3.3

Evaluate the following mentally:

1. $4 \div 1 =$
2. $9 \div 1 =$
3. $16 \div 1 =$
4. $25 \div 1 =$
5. $36 \times 1 =$
6. $49 \times 1 =$
7. $64 \times 1 =$

8. $81 \times 1 =$

9. $100 \div 100 =$

10. $144 \div 144 =$

11. $121 \div 121 =$

12. $196 \div 196 =$

13. $169 \times 0 =$

14. $225 \times 0 =$

15. $16 \times 0 =$

16. $25 \times 0 =$

17. $0 \div 1 =$

18. $0 \div 12 =$

19. $0 \div 9 =$

20. $0 \div 99 =$

21. $99 \div (22 - 22) =$

22. $161 \div (512 - 512) =$

23. $36 \div 0 =$

24. $81 \div 0 =$

25. $100 \div 0 =$

26. $64 \div 0 =$

27. $10^0 =$

28. $y^0 =$

29. $p^0 =$

30. $55^0 =$

Mental Exercise 3.4: Word Problems

1. A computer produces an error message #DIV/0! under certain conditions when you are using an Excel spreadsheet. Can you interpret this message in your own words?
2. If you multiply a number by 15 and then divide it by 15 as well, did you change the answer?
3. Can you provide an estimate of 11 divided by 0.000000000000001?

ESTIMATING

Estimating is a powerful and easy-to-use technique for simplifying complex-looking operations and expressions. Also, it is a quick way to know if your tentative answer (estimate) is "in the ball park"—that is, is it within a reasonable range to be correct.

For example, if you want to add 99 and 99, it is easy to round these numbers up to 100 and 100. Then add the two numbers (100 + 100 = 200), which is easy to do. Similarly, if the two numbers are 81 and 89, it is easy to round 81 down to 80 and 89 up to 90, and then add the two new numbers (80 + 90 = 170). You are within range of being correct. The same estimating process works for multiplication, subtraction, and division as well.

Mental Exercise 3.5

Evaluate the following mentally by estimating. Provide only the estimated answers; exact answers are not required for this exercise.

1. $19 + 19 =$
2. $18 + 18 =$
3. $89 + 89 =$
4. $101 + 102 =$
5. $29 - 19 =$
6. $49 - 18 =$
7. $99 - 39 =$
8. $199 - 99 =$
9. $18 \times 18 =$
10. $11 \times 18 =$
11. $21 \times 19 =$
12. $19 \times 32 =$

MENTAL CHAPTER REVIEW

Students need to perform the arithmetic operations below mentally. You can also choose to prepare a handout for these problems to provide clarity for some students if you feel they have a need.

Squares and square roots

1. $3^2 =$
2. $4^2 =$
3. $6^2 =$
4. $5^2 =$
5. $9^2 =$
6. $8^2 =$
7. $\sqrt{4} =$
8. $\sqrt{9} =$
9. $\sqrt{16} =$
10. $\sqrt{25} =$
11. $\sqrt{36} =$
12. $\sqrt{49} =$

Properties of 0

13. $5 + 0 =$

14. $11 - 0 =$

15. $6 \times 0 =$

16. $22 \times 0 =$

17. $\dfrac{8}{0} =$

18. $\dfrac{67}{0} =$

Properties of 1

19. $23 \times 1 =$

20. $83 \times 1 =$

21. $\dfrac{14}{1} =$

Addition with two terms

22. $4 + 12 =$
23. $5 + 13 =$
24. $11 + 11 =$

25. $7 + 6 =$
26. $9 + 8 =$
27. $11 + 9 =$

28. $8 + 2 =$
29. $12 + 18 =$
30. $16 + 4 =$

Subtraction with two terms

31. $8 - 6 =$
32. $11 - 6 =$
33. $9 - 2 =$

34. $14 - 12 =$
35. $28 - 22 =$
36. $12 - 5 =$

37. $11 - 4 =$
38. $9 - 1 =$
39. $8 - 2 =$

Addition and subtraction with multiple terms

    Students need to group the positive and negative terms first before doing the addition or subtraction operations. For example, in a problem like $11 - 5 + 3 - 4$, group 11 and 3 first. Next, group $- 5$ and $- 4$, which gives $11 + 3 - 5 - 4$. Add the positive terms first, and then add the negative terms to give $14 - 9$. This process gives a final answer of 5. If students show any signs of difficulty dealing with the addition of two negative numbers, teach those concepts here as well. You do not have to wait until later.

40. $5 + 2 + 1 =$

41. $6 + 3 + 1 =$

42. $6 + 3 + 2 =$

43. $4 + 2 + 4 =$

44. $5 + 1 + 3 =$

45. $5 - 2 - 2 =$

46. $10 - 4 - 6 =$

47. $7 - 2 - 5 =$

48. $8 - 2 - 1 =$

49. $9 - 1 - 7 =$

50. $5 + 1 - 6 =$
51. $6 + 2 - 7 =$
52. $5 - 3 + 2 - 4 =$
53. $5 - 4 + 2 - 3 =$
54. $6 + 1 - 2 - 4 =$

55. $-2 - 3 + 5 =$
56. $-5 - 4 + 9 =$
57. $-3 + 3 - 5 + 5 =$
58. $-6 + 2 + 5 - 1 =$
59. $-7 + 2 - 1 + 7 =$

More on the properties of 0 and 1

60. $4 \div 1 =$
61. $9 \div 1 =$
62. $16 \div 1 =$
63. $100 \div 100 =$

64. $144 \div 144 =$
65. $121 \div 121 =$
66. $0 \div 1 =$
67. $0 \div 12 =$

68. $0 \div 9 =$
69. $10^0 =$
70. $y^0 =$
71. $(249)^0 =$

Estimating answers: Exact answers are not required.

72. $21 + 19 =$
73. $18 + 18 =$
74. $89 + 89 =$

75. $29 - 19 =$
76. $49 - 18 =$
77. $99 - 39 =$

78. $18 \times 18 =$
79. $17 \times 17 =$
80. $21 \times 22 =$

MULTIPLICATION TABLES

Table 3.4 Multiplication Table 1

Student Name _____

MULTIPLICATION TABLE

	2	3	4	5	6	7	8	9	10
2									
3									
4									
5									
6									
7									
8									
9									
10									

Table 3.5 Multiplication Table 2

Student Name _____

MULTIPLICATION TABLE

	2	3	4	5	6	7	8	9	10	11
2										
3										
4										
5										
6										
7										
8										
9										
10										
11										

SAMPLE EXIT QUESTIONS

Simplify.
1. $(11^0 + 1^0 - 2)$
2. $(9 - 8)^0 * (z - 2)^0$
3. $\dfrac{(z - 5)*(123)*(a - b)}{(p - p)}$
4. Write 1/8 in decimal notation
5. Write 3/8 in decimal notation

SAMPLE PRIMING HOMEWORK

Simplify.

1. $\dfrac{(2300)*(2000)}{(200,000)}$
2. $\sqrt{64}$
3. $\sqrt[3]{64}$
4. $\sqrt{y^2}$
5. $\sqrt[3]{z^3}$
6. $\sqrt{64p^4}$

POINTS TO REMEMBER AND REVIEW

- Any number and/or term multiplied by 0 is always 0.
- You cannot divide by 0; it gives an undefined answer.
- An answer of 1 can also be written as $\frac{6}{6}, \frac{21}{21}, \frac{512}{512}$, and so on.
- You can either divide or multiply by 1 without changing the final answer.
- Recognize that 1/4 is 0.25, 1/2 is 0.5, and 3/4 is 0.75. Students need to overlearn these facts after initial understanding.
- Review and practice until overlearned.

Reminder: The homework, the sample tests, and the answer key for this chapter are provided in the booklet. For the rest of the chapters, the same pattern is followed.

4

Integers and Arithmetic Operations

Every time we turn around, we are dealing with numbers large and small—10, 1,000, 1,000,000, and so on. Sometimes, the size of the number makes people's minds go blank; it's just too big, like 1,000,000,000,000. Additionally, many people, adults as well as students, have problems with the processes of addition, subtraction, multiplication, and division of more than two-digit numbers. Then, to top it off, along come squares, square roots, and fractions. It's now mental shutdown time. But it shouldn't be.

To aid students in their mastery of these processes, this chapter involves the basics of addition, subtraction, multiplication, division, squares, and square roots. The students need to perform these operations mentally without the help of a calculator. As the teacher, instruct the students first in these basic operations, and then ask them to provide answers for the list of problems provided at the end of each section.

Be sure to ask each student by name to answer the questions based on their differential knowledge bases, discussed earlier in chapter 1. Do not allow the students to call out the answers. If you do, only the usual suspects in the front row will respond, and you will have lost the rest of the class.

Included in this chapter are mathematical operations like squares and square roots of simple numbers, as well as fractions and their mathematical operations.

DEFINITION OF INTEGERS

The lowest nonnegative integer is 0, and there is no largest integer. A typical set or a collection of integers is
$$\ldots, -6, -5, -4, -3, -2, -1, 0, 1, 2, 3, 4, 5, \ldots$$
This set can go on indefinitely.

Table 4.1 is provided to expose students early to these familiar words in the world of math and sciences. In a week or so, students should be expected to commit to memory these expressions and their math meanings.

Table 4.1 Useful Large and Small Numbers

Number	Word Name	Your Experience in 2007/2008
1	One	A soda can from a vending machine can cost $1.25
10	Ten	You could pay $30-40 for a full tank of regular gasoline
100	One hundred	Sneakers can cost between $90-$150
1000	One thousand	You can earn $1000 in four weeks by working a job that pays $6.25 per hour
1,000,000	One million	You can store 1.44 million characters (1.44 MB) on a 3 ½" floppy disk; a 500 page English book can be stored on one floppy disk
1,000,000,000	One billion	A CD can store 700 million bytes (700 MB, 0.7 GB); a collection of about 600 text-only books on one CD
1,000,000,000,000	One trillion	The US Federal budget in 2007 is about two trillion dollars
$\frac{1}{10} = 0.1$	One-tenth	A dime in US currency
$\frac{1}{100} = 0.01$	One-hundredth	A penny in US currency
$\frac{1}{1000} = 0.001$	One-thousandth or a milli	1/10th of one percent
$\frac{1}{1,000,000} = 0.000001$	One-millionth or a micro	Better than your chances of winning a state lottery

Example 4.1. Writing numbers into word names

    133 = One hundred thirty-three
    1,378 = One thousand, three hundred seventy-eight
    23,598 = Twenty-three thousand, five hundred ninety-eight
    34,011 = Thirty-four thousand, eleven
    456,734 = Four hundred fifty-six thousand, seven hundred thirty-four
    3,378,252 = Three million, three hundred seventy-eight thousand, two hundred fifty-two

Example 4.2. Writing word names into numbers

    Three hundred fifty-two = 352
    Five thousand, four hundred sixty-three = 5,463
    Forty-five thousand, six hundred thirty-two = 45,632
    Seventy-one thousand, nineteen = 71,019
    Three hundred twenty-three thousand, five hundred three = 323,503
    Six million, four hundred thirty-six thousand, one hundred forty-nine = 6,436,149

Exercise Set 4.1

Write a word name.

1. 213
2. 4,512
3. 7,012
4. 9,006
5. 31,568
6. 76,607
7. 327,576
8. 412,106
9. 306,006
10. 2,346,821
11. 6,412,703
12. 28,713,016
13. 76,031,064

Write a number.

1. Twenty-three
2. Four hundred twenty-one
3. Nine hundred six
4. Three thousand, eight hundred twelve
5. Eighty-one thousand, five hundred seven
6. Forty-one thousand, seven

7. Two hundred thirty-seven thousand, six hundred forty-five
8. Four hundred sixty-three thousand, three hundred four
9. Nine hundred thousand, six hundred forty-three
10. Six million, seven hundred thirty-two thousand, eight hundred fifty-three
11. Twenty-eight million, six hundred thousand, seven hundred thirty-four
12. Seventy-four million, three thousand, three
13. Twelve billion dollars
14. Six hundred million dollars
15. Twenty-three trillion dollars

Write a word name for the number in each sentence.

1. The U.S. population as of January 20, 2004, was approximately 292,417,056.
2. The world population estimate as of January 20, 2004, was 6,343,319,060.
3. The land area of the United States is approximately 9,166,660 square kilometers.

ADDITION OF WHOLE NUMBERS

Addition of whole numbers means combining or putting together the numbers involved.

Example 4.3. Adding Integers

Sample Example 4.3 Add 432 + 364

Step 1. Place values lined up in columns starting from the right hand side.

Step 2. Add numbers in the unit positions first

Step 3. Add numbers in the tens and the hundred positions

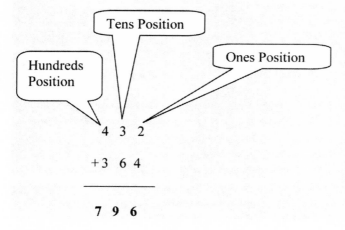

Example 4.4. Adding Integers Using the Carryover Technique

Sample Example 4.4 Add 5768 + 6475 using the *carryover* technique

Step 1. Place values lined up in columns starting from the right hand side.

Step 2. Add numbers 8 and 5 in the unit positions first. You get 13. Write 3 in the unit position of the answer and mentally carry over 1 in the tens position.

Step 3. Add numbers in the tens position 1, 6 and 7. The number one is a carry over from the unit position. The answer is 14. Write 4 in the tens position of the answer and mentally carry over 1 again in the hundreds position.

Step 4. Repeat this procedure for the hundredth and thousands positions.

$$
\begin{array}{r}
1\ 1\ 1 \ \leftarrow Carryovers \\
5\ 7\ 6\ 8 \\
+\ 6\ 4\ 7\ 5 \\
\hline
12\ 2\ 4\ 3
\end{array}
$$

Example 4.5. Adding Integers Using the Carryover Technique, continued …

Sample Example 4.5 Add 64 + 256

Step 1. Place values lined up in columns starting from the right hand side.

Step 2. Add numbers in the unit positions first. You get 10. Write 0 in the unit position of the answer and mentally carry over 1 in the tens position.

Step 3. Similarly, add numbers in the tens and hundred positions using the *carrying over* technique.

$$
\begin{array}{r}
6\ 4 \\
+\ 2\ 5\ 6 \\
\hline
3\ 2\ 0
\end{array}
$$

Exercise Set 4.2

Mental Addition

1. $3 + 4 =$
2. $5 + 4 =$
3. $6 + 5 =$
4. $7 + 6 =$
5. $8 + 7 =$
6. $9 + 8 =$
7. $8 + 8 =$
8. $9 + 9 =$
9. $7 + 7 =$
10. $6 + 6 =$
11. $11 + 11 =$
12. $12 + 12 =$
13. $13 + 13 =$
14. $14 + 14 =$
15. $16 + 16 =$
16. $12 + 13 =$
17. $13 + 15 =$
18. $14 + 17 =$
19. $15 + 19 =$
20. $16 + 20 =$

Add the following.
Allow students to use paper and pencil to complete this set of problems.

1. $23 + 67$
2. $49 + 51$
3. $19 + 21$
4. $28 + 42$
5. $15 + 315$
6. $201 + 63$
7. $32 + 512$
8. $512 + 512$
9. $346 + 469$
10. $589 + 234$
11. $1024 + 1024$
12. $2048 + 2048$
13. $3167 + 6867$
14. $4193 + 5438$
15. $9216 + 8195$
16. $8427 + 9218$
17. $9821 + 9538$
18. $2547 + 3538$
19. $6537 + 6537$
20. $7654 + 4567$

Word Problems

As a teacher, emphasize the words that imply an arithmetic operation and what is being asked of the students.

1. How much money will you have in the bank at the end of four weeks if you deposit $250 each week?
2. How much did you spend on shopping at the mall if you spent $10 on a movie, paid $6 for popcorn and Coke, and bought shoes for $84?
3. Your parents have been depositing $80 every month in your college account. How much did they deposit for you in one year?

SUBTRACTION OF WHOLE NUMBERS

Subtracting whole numbers means *taking away* a smaller number from a larger number or simply the difference between two numbers.

Example 4.6. Subtracting Integers

Sample Example 4.6 Subtract 596 - 252

Step 1. Place values lined up in columns starting from the right hand side.

Step 2. Subtract numbers in the unit positions first

Step 3. Subtract numbers in the tens and hundred positions

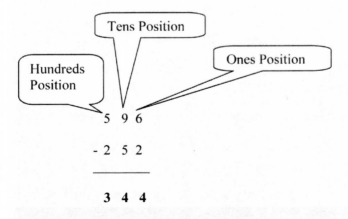

Example 4.7. Subtracting Integers using the Borrowing Technique

Sample Example 4.7 Subtract 7657 - 3269 using the *borrowing* technique

Step 1. Place values lined up in columns starting from the right hand side

Step 2. You cannot subtract 9 ones from 7 ones, but you can subtract 9 ones from 17 ones. You borrow 1 ten from 5 tens to get 17 ones. Now subtract 9 ones from 17 and write the answer under the unit position.

Step 3. You cannot subtract 6 tens from 4 tens (1 ten was borrowed from 5 tens leaving behind 4 tens), but you can subtract 6 tens from 14 tens. You borrow 1 hundred from 6 hundreds to get 14 tens. Now subtract 6 tens from 14 tens and write the answer under the tens position.

Step 4. Repeat this procedure for the hundreds and thousands positions.

```
    5 14 17  ← after borrowing from the higher position
  7  6  5  7
-  3  2  6  9
  ─────────────
  4  3  8  8
```

Example 4.8. Subtracting Integers using the Borrowing Technique, continued …

Sample Example 4.8 Subtract 64 from 256

Step 1. Place values lined up in columns starting from the right hand side.

Step 2. Subtract numbers in the unit positions first. You get 2. Write 2 in the unit position of the answer.

Step 3. You cannot subtract 6 ones from 5 ones, but you can subtract 6 tens from 15 tens. You <u>borrow</u> 1 hundred from 2 hundreds to get 15 tens. Now, subtract 6 tens from 15 tens, and write the answer in the tens position.

Step 4. Subtract 0 hundreds from 1 hundreds and write the answer under the hundreds position.

```
     2  5  6
  -     6  4
     ────────
     1  9  2
```

Exercise Set 4.3

Mental Subtraction

1. 9 − 4 =
2. 8 − 4 =
3. 8 − 5 =
4. 9 − 6 =
5. 9 − 7 =
6. 9 − 6 =
7. 8 − 8 =
8. 9 − 9 =
9. 7 − 7 =
10. 9 − 6 =
11. 13 − 11 =
12. 17 − 12 =
13. 19 − 13 =
14. 19 − 14 =
15. 21 − 16 =
16. 28 − 13 =
17. 29 − 15 =
18. 28 − 17 =
19. 27 − 19 =
20. 26 − 20 =

Subtract the following.

Allow students to use paper and pencil to complete this set of problems.

1. 67 − 23 =
2. 51 − 29 =
3. 19 − 21 =
4. 42 − 28 =
5. 315 − 15 =
6. 201 − 63 =
7. 512 − 32 =
8. 512 − 512 =
9. 469 − 346 =
10. 589 − 234 =
11. 2048 − 1024 =
12. 4096 − 2048 =
13. 6867 − 3167 =
14. 5438 − 4193 =
15. 9216 − 8195 =
16. 9427 − 9218 =
17. 9821 − 9538 =
18. 3538 − 2547 =
19. 6537 − 3437 =
20. 7654 − 4567 =

Word Problems

Emphasize the words that imply an arithmetic operation and what is being asked of the students. Sometimes, a problem has more than one operation involved. Make sure the students understand that.

1. How much money will you have in the bank at the end of four weeks if you deposited $250 each week and then withdrew $900?
2. You had $100 with you when you went to the mall. How much is left if you spent $10 on a movie, $6 on popcorn and coke, and $84 for shoes?
3. Your parents have been depositing $80 every month in your expense account. You took out $40 every month for your personal expenses and gas. How much will you have at the end of one year?

MULTIPLICATION OF WHOLE NUMBERS

We will restrict ourselves here to the multiplication of whole numbers in the range of 1–20. Your students must have mastered the multiplication tables by now. For multiplication involving large whole numbers, a calculator can be used, but they do not need one here.

Multiplying 4 * 5 can mean, for example, having 4 five-dollar bills amounting to a total of $20. Similarly, 5 * 2 can mean having 5 two-dollar bills amounting to a total of $10. Throughout the book, the multiplication symbols will be *or ×.

Exercise Set 4.4

Mental Multiplication

1. $6 \times 4 =$

2. $8 \times 4 =$

3. $8 \times 5 =$

4. $9 \times 6 =$

5. $8 \times 7 =$

6. $5 \times 6 =$

7. $8 \times 8 =$

8. $9 \times 9 =$

9. $7 \times 7 =$

10. $9 \times 6 =$

11. $10 \times 11 =$

12. $10 \times 12 =$

13. $13 \times 13 =$

14. $14 \times 14 =$

15. $16 \times 16 =$

16. $15 \times 15 =$

17. $16 \times 15 =$

18. $12 \times 11 =$

19. $12 \times 12 =$

20. $13 \times 9 =$

21. $5 \times 7 =$

22. $5 \times 8 =$

23. $5 \times 9 =$

24. $5 \times 10 =$

25. $4 \times 6 =$

26. $4 \times 7 =$

27. $4 \times 8 =$

28. $4 \times 9 =$

29. 6 × 7 =

30. 6 × 8 =

31. 6 × 9 =

32. 6 × 10 =

33. 7 × 4 =

34. 7 × 5 =

35. 7 × 6 =

36. 7 × 7 =

37. 11 × 11 =

38. 11 × 13 =

39. 11 × 14 =

Word Problems

Again, emphasize the words that imply an arithmetic operation and what is being asked of the students.

1. How much did the class donate to the children's fund if each student in a class of 20 donated $3.00?
2. How many total computers did Apple Computers donate to your school system of 40 schools if each school received four Apple computers?
3. Your school has four sections of senior high school, and each section has a maximum capacity of 20 students. How many students at the most can you enroll in the school?

DIVISION OF WHOLE NUMBERS

Division of 100 by 10 can mean, for example, getting change in $10 bills for a $100 note. The answer is 10. Similarly, 20 ÷ 5 can mean getting four notes of $5 for a $20 bill.

Exercise Set 4.5

Mental Division

1. 6 / 3 =
2. 8 / 4 =
3. 10 / 5 =
4. 12 / 6 =
5. 14 / 7 =
6. 9 / 3 =
7. 16 / 8 =
8. 18 / 9 =
9. 7 / 7 =
10. 12 / 6 =
11. 10 / 5 =
12. 10 / 2 =
13. 14 / 7 =
14. 21 / 3 =
15. 16 / 4 =
16. 15 / 3 =
17. 45 / 15 =
18. 33 / 11 =
19. 48 / 12 =
20. 36 / 9 =
21. 49 / 7 =
22. 48 / 8 =
23. 63 / 9 =
24. 90 / 10 =
25. 54 / 6 =
26. 56 / 7 =
27. 72 / 8 =
28. 81 / 9 =
29. 42 / 7 =
30. 64 / 8 =
31. 45 / 9 =
32. 70 / 10 =
33. 32 / 4 =
34. 35 / 5 =
35. 30 / 6 =
36. 35 / 7 =
37. 88 / 11 =
38. 52 / 13 =
39. 56 / 14 =
40. 75 / 15 =

Word Problems

Stress the words that imply an arithmetic operation and what is being asked of the students.

1. You went to your local bank to withdraw $500 and asked the cashier to give you only fifty-dollar bills. How many $50 bills are you expecting to receive from the cashier?
2. In a supermarket, you paid $60 to buy six cases of a soft drink. How much did it cost you for one case?
3. You went to purchase some DVDs from a movie store. How many DVDs can you buy for $45 if the price of one DVD is $5?

Integers and Arithmetic Operations

SQUARES OF WHOLE NUMBERS

Squaring a number means multiplying the number by itself. For example, the square of 5 is 25, that is, 5 times 5. The square of 9, multiplying 9 by itself, 9×9, is 81. When we want to write this task in a short form, we write these expressions as 5^2 and 9^2. These squares are usually read as five squared, or five raised to the power of 2, and 9 squared or raised to the power of 2.

Exercise Set 4.6

Square mentally:

1. $2^2 =$
2. $3^2 =$
3. $4^2 =$
4. $5^2 =$
5. $6^2 =$
6. $7^2 =$
7. $8^2 =$
8. $9^2 =$
9. $10^2 =$
10. $11^2 =$
11. $12^2 =$
12. $13^2 =$
13. $14^2 =$
14. $15^2 =$
15. $14^2 =$
16. $13^2 =$
17. $12^2 =$
18. $11^2 =$
19. $10^2 =$
20. $9^2 =$
21. $8^2 =$
22. $7^2 =$
23. $6^2 =$
24. $5^2 =$
25. $4^2 =$
26. $3^2 =$
27. $2^2 =$
28. $3^2 =$
29. $5^2 =$
30. $6^2 =$
31. $7^2 =$
32. $8^2 =$
33. $9^2 =$
34. $20^2 =$
35. $30^2 =$
36. $4^2 =$
37. $6^2 =$
38. $7^2 =$
39. $8^2 =$
40. $9^2 =$

Word Problems

Emphasize the words that signal an arithmetic operation and what is being asked of the student. Point out that more than one arithmetic operation may be involved.

1. The floor surface (area) is measured in square feet, and a carpenter will multiply the length by the width of a room to obtain the measurements of floor surface. If your room is a square of 12 feet in length, what will be the floor surface in ft^2? At a minimum, how many tiles will you need to buy for this room if the area of a tile is one square foot?
2. In your home, one side of a square table is 3 feet long. What is the surface area of the table in ft^2?

EASY SQUARE ROOTS OF WHOLE NUMBERS

If a number is a product of two *identical* factors, for example, 3 × 3 or 6 × 6, then either factor is termed the square root of that number. For example, the square root of 25 is 5 because 25 = 5 × 5. The standard square root sign is $\sqrt{\ }$; it is also called a *radical sign*. Sometimes, the square root sign is also written as ½ in the superscript (exponent). For example, the expressions $\sqrt{64}$ and $(64)^{1/2}$ are equivalent, and the answer is 8 in both cases.

Exercise Set 4.7

Finding Square Roots Mentally

1. $\sqrt{4} =$
2. $\sqrt{9} =$
3. $\sqrt{16} =$
4. $\sqrt{25} =$
5. $\sqrt{36} =$
6. $\sqrt{49} =$
7. $\sqrt{64} =$
8. $\sqrt{81} =$

9. $\sqrt{100} =$

10. $\sqrt{144} =$

11. $\sqrt{121} =$

12. $\sqrt{196} =$

13. $\sqrt{169} =$

14. $\sqrt{225} =$

15. $\sqrt{16} =$

16. $\sqrt{25} =$

17. $\sqrt{36} =$

18. $\sqrt{81} =$

19. $\sqrt{100} =$

20. $\sqrt{4} =$

21. $\sqrt{400} =$

22. $\sqrt{900} =$

23. $\sqrt{1600} =$

24. $\sqrt{3600} =$

25. $\sqrt{2500} =$

26. $\sqrt{4900} =$

27. $\sqrt{8100} =$

28. $\sqrt{6400} =$

29. $\sqrt{10,000} =$

30. $\sqrt{1,000,000} =$

Word Problems

Once again, point out the words that signal an arithmetic operation and what is being asked of the students.

1. The floor area of a square room in your house is 169 square feet. What is the length and width of the room in feet?
2. What is square root of a square? i.e., $\sqrt{20^2}$
3. What is the answer for square root of $\sqrt{x^2}$?
4. Your father told you to buy a square sheet of plastic of 144 square feet. What is your estimate for the length in feet of one side of this sheet?

REVIEW OF FACTS ABOUT 1 AND 0

1. Any number divided by 1 is the same number, e.g., 5 / 1 = 5
2. Any number multiplied by 1 is the same number, e.g., 9 × 1 = 9
3. Any nonzero number divided by itself is 1, e.g., 7 / 7 = 1
4. Any number multiplied by 0 is 0, e.g., 21 × 0 = 0
5. Zero divided by a nonzero number is 0, e.g., 0 / 6 = 0
6. A number can't be divided by zero. If you try to divide any number by zero, the result is undefined. Sometimes, in the sciences, we define it as infinity (∞).
7. Any number or variable raised to the power of 0 is 1, e.g., $(10)^0 = 1$; $(555)^0 = 1$; $n^0 = 1$.

Exercise Set 4.8

Evaluate the following mentally.

1. 4 / 1 =
2. 9 / 1 =
3. 16 / 1 =
4. 25 / 1 =
5. 36 × 1 =
6. 49 × 1 =
7. 64 × 1 =
8. 81 × 1 =
9. 100 / 100 =
10. 144 / 144 =
11. 121 / 121 =
12. 196 / 196 =
13. 169 × 0 =
14. 225 × 0 =
15. 16 × 0 =
16. 25 × 0 =

Integers and Arithmetic Operations

17. 0 / 1 =
18. 0 / 12 =
19. 0 / 9 =
20. 0 / 99 =
21. 36 / 0 =
22. 81 / 0 =
23. 100 / 0 =
24. 64 / 0 =
25. $(23)^0 =$
26. $(89)^0 =$
27. $(512)^0 =$
28. $z^0 =$

Word Problems

Note the words that indicate an arithmetic operation, and have the students state what is being asked of them.

1. A computer produces an error message, #DIV/0! given certain conditions when you are using an Excel spreadsheet. Can you interpret this message in your own words?
2. If you multiply a number by 15 and divide by 15, did you change the answer?
3. Can you provide an estimate of 21 divided by 0.000000000000001?
4. Simplify $(32)^0 * (1024)^0 * (Z)^0 =$
5. Simplify $(16)^0 (32) * (1024)^0 * (0) =$

MENTAL CHAPTER REVIEW

You, as the teacher of the course, should act only as a guide to review the problems in this section. Ask questions of every student based on his or her differential knowledge base to assure equal participation.

Writing a word name for a whole number

1. 23
2. 312
3. 256
4. 512
5. 1024
6. 2046
7. 4096
8. 25

9. 50
10. 100
11. 150
12. 200
13. 18
14. 27
15. 36
16. 45
17. 54
18. 63
19. 126
20. 72

Writing a number for a word

1. Twenty-four
2. Four hundred twenty-three
3. Nine hundred nine
4. Three thousand, five hundred six
5. Thirty-two thousand, eight

MULTIPLICATION, ADDITION, SUBTRACTION, AND DIVISION OF WHOLE NUMBERS

Perform the indicated operations.

1. $3 \times 11 =$
2. $4 \times 12 =$
3. $2 \times 10 =$
4. $5 \times 7 =$
5. $6 \times 7 =$
6. $7 \times 8 =$
7. $8 \times 9 =$

8. $9 \times 9 =$
9. $11 \times 11 =$
10. $12 \times 12 =$
11. $4 + 7 =$
12. $5 + 8 =$
13. $6 + 9 =$
14. $7 + 7 =$
15. $9 + 9 =$
16. $11 + 11 =$
17. $12 + 12 =$
18. $13 + 13 =$
19. $14 + 14 =$
20. $14 + 16 =$
21. $11 - 2 =$
22. $12 - 3 =$
23. $13 - 3 =$
24. $14 - 5 =$
25. $15 - 8 =$
26. $16 - 9 =$
27. $17 - 9 =$
28. $18 - 9 =$
29. $19 - 9 =$
30. $20 - 5 =$
31. $10 / 2 =$
32. $12 / 2 =$
33. $15 / 3 =$
34. $20 / 4 =$
35. $20 / 5 =$
36. $25 / 5 =$

37. 35 / 5 =
38. 40 / 8 =
39. 60 / 10 =
40. 72 / 12 =
41. 6^2 =
42. 7^2 =
43. 9^2 =
44. 10^2 =
45. 11^2 =
46. 13^2 =
47. 12^2 =
48. 15^2 =
49. $\sqrt{100}$ =
50. $\sqrt{9}$ =
51. $\sqrt{16}$ =
52. $\sqrt{25}$ =
53. $\sqrt{36}$ =
54. $\sqrt{49}$ =
55. $\sqrt{64}$ =
56. $\sqrt{81}$ =
57. $\sqrt{100}$ =
58. $\sqrt{144}$ =
59. $\sqrt{169}$ =
60. $\sqrt{121}$ =
61. 5 × 0 =
62. 10 × 0 =
63. 0 × 21 =

64. $9 + 0 =$

65. $23 + 0 =$

66. $14 - 0 =$

67. $28 - 0 =$

68. $10 / 0 =$

69. $5 / 0 =$

70. $122 / 0 =$

71. $0 / 22 =$

72. $0 / 36 =$

73. $(10)^0 =$

74. $(1000)^0 =$

75. $(xy)^0 =$

MULTIPLICATION TABLES

Table 4.2 Multiplication Table 1

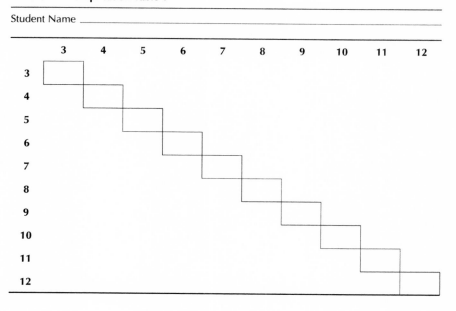

Table 4.3 Multiplication Table 2

Student Name _____

	4	5	6	7	8	9	10	11	12	13
3										
4										
5										
6										
7										
8										
9										
10										
11										
12										
13										

SAMPLE EXIT QUESTIONS

Simplify:

1. $[9^0 + 1^0 + (y - z)^0 + 7]$
2. $\dfrac{11 * 1100 * 231}{23,100}$
3. $\sqrt{169}$
4. $\sqrt{100}$
5. $\sqrt[3]{125}$

SAMPLE PRIMING HOMEWORK

1. What is the mean of 11, 19, 5, and 25?
2. What is the median of 11, 35, 40, 8, 22, 30, and 38?
3. What is the median of 11, 35, 40, 8, 22, 30, 38, and 32?
4. What is the mode of 11, 22, 33, 44, 55, 44, 66, and 44?

POINTS TO REMEMBER AND REVIEW

- Simplify terms in the numerator and denominator before multiplying and dividing, e.g., $\frac{9*1300*200}{260,000} = 9$
  - ✓ First, cancel four zeros in the numerator and denominator.
  - ✓ Second, multiply 13 and 2, giving 26.
  - ✓ Cancel 26 in the numerator with 26 in the denominator.

- Cancel the follow-up zeros in the denominator and numerator terms for whole numbers.
- The square root of a term that is squared is the term itself, e.g., $\sqrt{z^2} = z$.
- The square root of a term that has an even number of zeros, following a 1 in the radical sign, is 1 followed by half the number of zeros, e.g., $\sqrt{100} = 10$; $\sqrt{10,000} = 100$ (the radical sign has either two or four zeros, whereas the answer has half the number of zeros, e.g., one or two).

5

Data, Statistics, and Graphs

We know many students dislike the words "graphs" and "statistics" because they often signal trouble in interpreting what they mean, so they avoid them. But wait: don't let the students entice you into feeling the same way. These words actually signal topics that are some of the easier and more useful ones for students to deal with in both their academic and daily living experiences. Examples of statistics are all around us, and we see graphs all day long. What the statistics and the graphs are showing us are different types of data or specific facts presented in numbers. By the way, the word *data* is the *plural* form of *datum* (both straight from Latin).

You see data all around you. The basketball or football scores on the scoreboard, scores on your tests, winter or summer temperatures, or different dress or pant sizes are all examples of data. All of these data can be presented in tables, as statistical measures, or in graphs. We will be dealing with these various ways of organizing data in this chapter in order to see their main characteristics.

COLLECTING CLASS DATA IN A SET

We have used an inductive approach for this section of the chapter, meaning that we will derive definitions from students' experiences. Below are weight data in pounds taken from students in our classes, rounded to the nearest multiple of 10. Using multiples of 10 keeps the addition and division of numbers simple. We have collected these data for sets of four, five, six, and ten students:

Four students: 100, 160, 140, 200 lb
Five students: 100, 180, 120, 140, 160 lb
Six students: 100, 180, 120, 150, 160, 130 lb
Ten students: 100, 180, 120, 150, 160, 120, 150, 150, 100, 150 lb

THE ARITHMETIC MEAN

Scenario 1. Ask a student, individually by name, "What is the *average* of the weights of the four students?" If the student doesn't understand the term *average* or is not sure how to proceed, ask another student by name that may have raised his or her hand, "Can you help us out here?" Quite likely, the second student will produce the right answer of 150 lb.

If neither student solves the problem, stop right there. Don't prolong the agony. Instead, elicit from them what their understandings of *average* are. They may verbalize such statements as "in the middle," "not at the top or the bottom," or "like getting a C in a course." In short, they are displaying a grasp of average in the sense you are seeking, and you need to let them know they are doing just fine.

Then you may want to try again and see if they can compute the average or mean for this data set. Be sure to point out that 150 pounds is indeed hovering around the middle of the set, sometimes called a *distribution*. Then have the students find the mean for each of the other sets.

Scenario 2. If a student immediately answers correctly, ask the student by name to show you and the other students exactly how the answer was computed. Use the explanation given by the student to introduce the formula for this average, that is, the arithmetic mean.

$$\text{Mean or Average} = \frac{\text{Sum of the numbers in the set}}{\text{Number of data items in the set}} = \frac{x}{n}$$

Then compare the correct student solution of the problem to the definition and introduce the formula for the arithmetic mean. Also point out that the *average* or *mean* for a set doesn't have to be an actual score or observation, although it often is. Have the students find the mean or average for each of the other sets.

THE MEDIAN

Scenario 1. Explain to the students, "We are now going to find another average. It's called the *median*." Ask a student, again by name, "Can you find the median for the set having five students?" If the student answers correctly (median = 140 lb), ask that student to explain the answer to the others. Proceed to the example involving six numbers, and ask someone else for the median for this set of weights.

Scenario 2. If the first student does not answer correctly, ask another student, by name, to explain his or her understanding of the word *median*. You may receive such answers as "the middle strip dividing lanes on a highway" or "half of a loaf of bread." Again, the students have demonstrated a grasp of the concept. Explain immediately that the median is the exact middle score in a data set. Then ask again, "What is the exact middle score for the data set having five students?" This time, someone should come up with the answer of 140 pounds.

Now, call on a different student by name: "What is the median for the weights of six students? Remember the median is exactly in the middle when the data set is arranged in either an ascending or a descending order." Hopefully, someone can arrive at the answer of 140 lb. Then repeat this process with the data set having the weights of ten students.

> *Median* is defined as the middle number of the set if the set contains an odd number of items when these data are arranged in either a descending or ascending order. If the set contains an even number of items, the median is the average of the two middle numbers. For example, when the weights of five students are 100, 180, 120, 140, 160 lbs, the median will be 140 because 140 is the middle weight when these items are arranged in the ascending order. However, when the weights of six students, which is an even set, are 100, 180, 120, 150, 160, 130 lbs, the two middle items of the set when arranged in the ascending order are 130 and 150. Their average is 140 lbs, which is then the median of this set.

THE MODE

Now you tell the group, "There is still another average. Does anyone know what it is called? How about you?" selecting one student. You continue: "We can find it usually by just examining the numbers in the data set, once it's been arranged in either ascending or descending order. Let's look at this set of ten students, which has been arranged in an ascending order."

Ten students: 100, 100, 120, 120, 150, 150, 150, 150, 160, 180 lb

With the data set properly arrayed, the students should easily see the correct answer for the mode of the weight of ten students as being 150 lb because 150 occurs most often in the data set.

> *Mode* is defined as the number or numbers that <u>occur most often in a data set</u>, the ones (scores or observations) with the <u>highest frequency</u>. A set of numbers can have more than one mode; sometimes, it will have no mode at all. For example, the set 3, 5, 6, 9, 11, 15 has no mode because each number occurs the same number of times, in this case, once. In contrast, the set of numbers 11, 24, 24, 24, 33, 40, 45, 45, 45, 56, 70, 96 has two modes, 24 and 45, because it has two numbers that occur most often and with equal frequency.

CONDITIONS FOR USING THE MEAN, THE MEDIAN, AND THE MODE

The three statistics we have just computed are also commonly described in statistics as *measures of central tendency*. What this fancy phrase is saying that, in general, the three measures tend to be located in the middle of the scores or observations of a data set. However, all three statistics cannot be computed for all data sets; sometimes there is no mode.

When a data set is made up of categories (e.g., political party affiliation with the categories of Republicans, Democrats, Independents, and Undecided), then we can do the counts within the categories and determine which party affiliation has the largest number of members (the mode), but we cannot legitimately compute the median or the mean. Either one becomes a meaningless computation in the case of data in categories. With categories, you cannot put an observation in more than one category. Ask your students for examples of other categories. Some common ones are ethnicity (e.g., African-American, white, Latino, Asian, Native American). Another is gender, having two categories: male and female.

Businesses use the mode all the time. For example, the mode is very useful for a store in stocking the most popular sizes of clothes or shoes, most fashionable colors of blouses, or most purchased brands of perfume. The underlying feature is that the data here are in categories. Ask your students to think of examples here as well. When a data set has more than one mode, only two modes usually are appropriate.

With the median and the mean, some key points must be kept in mind when using them to summarize key features of a data set and then draw conclusions. When the items in a data set have extreme values—for example, the value 30 in the data set of {30, 70, 75, 80, 85} is very low when compared with the other numbers in the set—it then becomes advisable to use the median rather than the mean to arrive at a more accurate description of the set.

In this particular instance, the average (mean) is 68, which doesn't really accurately describe or summarize the set. In contrast, the median for this set, 75 (the score in the middle), is a more accurate reflection of this particular data set. The presence of extreme scores at one end of a data set is known as *skewing*.

Feel free to use the mean when the values of items in a data set are close together or when the observations making up a particular data set are spread out about equally from the center of the distribution. Oftentimes, all three measures are used in presenting the information for a data set if the observations in the data set will allow them to be used.

Using all three simply presents a more accurate picture of the clustering points in a data set than any one of the measures alone. This practice is commonly used in research work in education, the social sciences, and the sciences. Typically, in the sciences, the data sets will make use of the mean, but all three measures usually can be used.

RANGE

One other statistical measure is presented here. This measure is used to make a quick estimate of how the scores or observations in a data set spread out from the mean or median. In some data sets, the observations are all bunched together. In others, they may be spread out. This measure of spread or dispersion is called the *range*, and it is written as, $Ra = (X_H - X_L) + 1$, where X_H is the highest value and X_L the lowest in the data set. (Usually, the subscripts H and L are in lower case letters; here, they are used in upper case because the lower case l looks like the number 1.)

Let us now return to the weights of our four students from the beginning of this section. These weights were: 100, 140, 160, and 200 lb. The range for these observations is: $(200 - 100) + 1 = 101$, which is quite large. Therefore, we can say that the data set has a wide range or is "spread out."

If we take the weight of five female cheerleaders, their weights might be 100, 110, 120, 120, and 130 lb. The range of these observations is: $(130-100) + 1 = 31$, which is a comparatively narrow range. This data set is bunched together.

COLLECTING CLASS DATA IN A TABLE FOR BAR AND PIE GRAPHS

Now, let us shift to another way of organizing numerical data. Table 5.1 shows a frequency distribution for the ethnicities of one of the author's classes. Tell the students that these data can be organized in several ways. We use *f* for the frequency in a data set. Provide your students with the template of the table. Then instruct them to organize the data in a tabular form in the following way by orally giving the frequencies/count for each category.

Table 5.1 A Count of Different Ethnicities in a Class

African-American f	White f	Hispanic f	Asian f
12	6	4	2

After the students have completed the table, tell them they also can organize data in different *graphical* ways. Ask the students if they have any ideas about what form these graphical ways can take—for example, bar graph, pie chart, line graph, or frequency polygon. If the students note the different types of graphs, give them the shells for the graphs with the labels for the x- and y-axes and have each student transform the frequencies in the table into a bar graph. Some of the students are going to get it right without any difficulty, but others may need assistance. Make certain everyone understands before moving on to another form of graphical presentation.

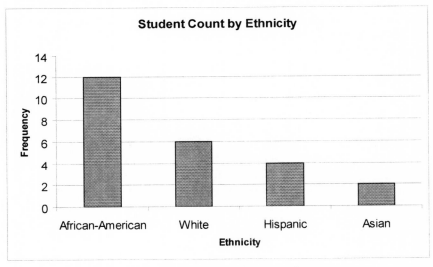

Graph 5.1 A Bar Graph of Class Ethnicity and its Frequency

To involve students in their learning, use this exercise to teach new concepts, build on what they already know, reinforce what they already might know, and identify gaps in their math knowledge by asking the following questions:

Example 5.1. Reading a Frequency Table

Question	Answer
How many Hispanic students are there in the class?	4
How many African-American students are there in the class?	12
How many more African-American students are there than Asians?	10
What is the ratio of White to African-American students?	$\frac{1}{2}$
What percent of the class is other than African-American?	50%

Now introduce the pie chart.

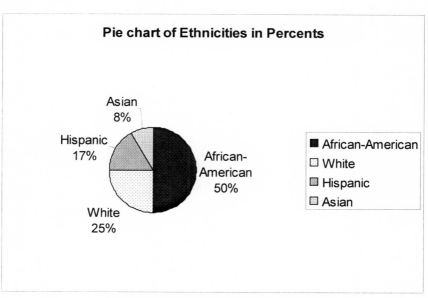

Graph 5.2 *A Percent Pie Graph of Class Ethnicity*

As before, involve students in providing answers to the following questions based on their "differential knowledge bases."

Example 5.2. Reading a Pie Chart

Question	Answer
What percent of students are Hispanics in the class?	17%
What percent of students are not African-Americans in the class?	50%
What is the total percent of students who are Whites and Asians?	33%
What is the ratio of White students to the total class?	$\frac{1}{4}$
What percent of students in the class is non-white?	75%

COLLECTING DATA IN A FREQUENCY DISTRIBUTION FOR A LINE GRAPH OR A FREQUENCY POLYGON

Example 5.3. Age versus Height of a Person

Age	Height
18	65
19	70
22	75
25	72
26	71
27	70
28	69
31	72
32	77
40	62

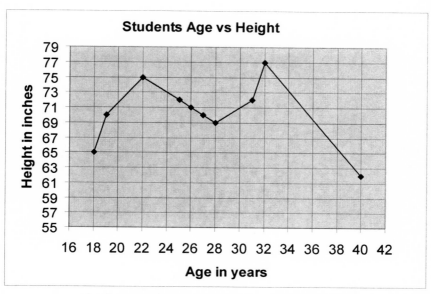

Graph 5.3 Students' Age in Years and Height in Inches

Example 5.4. Reading a Frequency Table

Question	Answer
What is the lowest height in this class?	62 inches
What is the age of the tallest student?	32 years
What is the height of a student who is 28 years old?	69 inches
What is the age of a student who is 75 inches tall?	22 years

MENTAL CHAPTER REVIEW

Find the mean or average.

1. 5, 6, 10
2. 10, 30, 50
3. 100, 400, 400
4. 15, 25, 35

Find the median.

5. 24, 34, 43, 56, 71
6. 129, 78, 56, 45, 32

7. 34, 56, 60, 70, 85, 92
8. 189, 197, 200, 400, 456, 734

Find the mode.

9. 22, 33, 22, 56, 89
10. 11, 13, 16, 21, 34, 56, 76
11. 15, 32, 56, 45, 32, 75, 32, 89
12. 123, 125, 167, 186, 167, 123, 167, 234, 123

For the following, find the mean (average), median, and any modes that exist. Also, compare the mean and the median to determine the statistic that is more suitable.

13. 2, 90, 110, 90, 8
14. 10, 50, 60, 80
15. 5, 10, 15, 20, 35, 35

The following frequency distribution table shows the cost of first-class postage in the United States starting in 1980. Use the table below to answer the three questions that follow.

Example 5.5. Cost of a First-class Postage Stamp during a 30-year Period

Year	First-class postage
1980	18¢
1983	20¢
1986	22¢
1989	25¢
1992	28¢
1995	32¢
1998	33¢
2001	35¢
2004	39¢
2007	41¢

16. What was the cost for first-class postage in 1989?
17. What is the increase in cents for the first-class postage from 1992 to 2007?
18. What was the year when the cost for first-class postage was 22 cents?

The following line graph shows the number of accidents in the U.S. per 100 drivers by age groups. Use this graph to answer the four questions that follow.

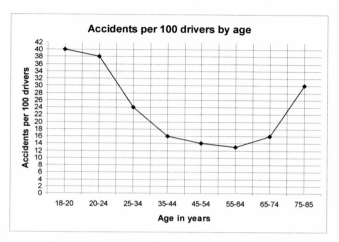

Graph 5.4 Number of Accidents in the U.S. Per 100 Drivers by Age Groups

19. Which group has the highest number of accidents per 100 drivers?
20. Which age group has the lowest number of accidents per 100 drivers?
21. Which age groups have the same number of accidents per 100 drivers?
22. For what age group does the number of accidents drop to 24 per 100 drivers?

The pie graph below shows the grade distribution in a typical math course. Use the data in this graph to answer questions that follow.

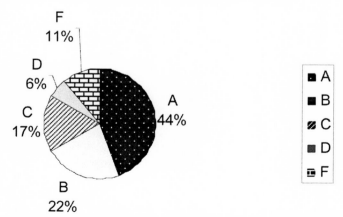

Graph 5.5 Grade Distribution in Percent for Typical Math Course

23. What percentage of the class earned an "A" in the course?
24. What percentage of students failed the course?
25. What percentage of students in the class who passed did not earn a grade of "A"?

CHALLENGING PROBLEM

The Fahrenheit temperatures in Washington, D.C., in a typical week in April are, as follows:

Table 5.2 Temperatures of a Typical April Week in D.C.

Mon	Tue	Wed	Thu	Fri	Sat	Sun
65	78	75	69	78	76	70

If m represents the median temperature, c represents the temperature that occurs most often (mode), and a represents the average (arithmetic mean) of the seven temperatures,

1. What is the average temperature expressed as $a = $ _____?
2. What is the median temperature expressed as $m = $ _____?
3. What is the mode expressed as $c = $ _____?
4. Express a, m, and c in terms of =, >, or < arithmetic relationships. (In this question, > is the *greater than* sign, whereas < is the *less than* sign.)

SAMPLE EXIT QUESTIONS

1. What is the median of 3, 11, 9, 5, 7, 13, and 15?
2. What is the median of 16, 30, 28, 26, 24, 18, 20, and 22?
3. What is the mode of 9, 22, 34, 55, 34, 45, and 34?
4. What is the range of 3, 11, 9, 5, 7, 13, and 15?

SAMPLE PRIMING HOMEWORK

1. The measurements of two angles of a triangle are 45° and 55°. What is the measurement of the third angle?
2. The measurement of one angle of a right angle triangle is 40°. What are the measurements of the other two angles?

3. The measurement of one angle of an isosceles triangle is 70°. What are the measurements of the other two angles?
4. What are the measurements of the three angles of an equilateral triangle?

POINTS TO REMEMBER AND REVIEW

- Arrange the items in a data set in an ascending order.
- In finding the median of a data set, having already arranged the items in ascending order, count the number of items in the data set. Median is defined as the *middle number* of the set if the set contains an *odd* number of items. If the set contains an *even* number of items, the median is the *average* of the two middle numbers.
- The mean alone should not be used if the data are skewed, that is, if there are extreme values at one end of a distribution. The median should also be shown.
- The range, the difference between the highest and the lowest terms of a data set, is a quick measure of how far the distribution spreads out from the center.

6

A Sometimes Love–Hate Relationship

Facts about Triangles, Intersecting Lines, and Units of Measurements

Many students are afraid of geometry, and their thinking about the subject stops right there—with the word. The result is that they try to avoid any further contact with the subject. However, the truth is that the facts about triangles, intersecting lines, and units of measurement are comparatively simple; these facts involve addition and subtraction of whole numbers. More importantly, they have applications and uses in the sciences. Students need to understand what is involved and then apply and practice these facts so they can do the operations mentally.

Students who manage to master these simple concepts will enjoy geometry, because it is really simple and elegant. The only exception to this may be the Pythagorean Theorem; it involves squares and square roots, and they could be somewhat challenging at the beginning. But the Pythagorean Theorem is really very elegant—and useful—as well.

Let us begin with the triangles.

TRIANGLES

Types of Triangles

Have the students draw and name the particular triangle. Ask them about the kinds of triangles they are familiar with and their properties. Many students have heard these names, and they know something about them. If the students do not know or remember the names of the triangles, then draw them on the chalkboard and see if that jogs somebody's memory a bit. Have a student who does remember come up to the board and label a particular triangle. Also, let that student describe the characteristics of that triangle. Repeat the process with other students. Then develop the matrix with students' help in terms of the types of triangles, their respective sides, and the angles.

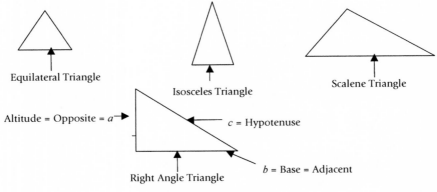

Image 6.1 Triangle Types

Characteristics of Triangles

Two numbers to remember are 90° for a right triangle and 180° when dealing with any triangle. In addition, students should commit to memory the following facts about triangles in general and the Pythagorean Theorem as applied to right triangles:

- Sum of all the angles of a triangle = 180°
- Pythagorean Theorem: $c = \sqrt{(a^2 + b^2)}$, where a, b, and c are the length of the opposite, adjacent, and hypotenuse sides of a right triangle

You should engage students when discussing the following characteristics of triangles:

Triangle type	Characteristics in terms of sides	Characteristics in terms of angles
Equilateral	All sides are equal	All angles are equal to 60°
Isosceles	Two sides are equal	Angles opposite to the sides are equal
Scalene	All sides are different	All angles are different
Right	Sides can be determined using the Pythagorean Theorem	One angle always equals 90°

You should also make up other mental problems to involve as many students as possible.

Manipulating the Pythagorean Theorem

Explain to the students the importance of these three "clues" in a problem: right triangle, an angle of 90°, or the Pythagorean Theorem. The problem is asking them to use the Pythagorean Theorem.

The equation for the Pythagorean Theorem is

$$c^2 = a^2 + b^2,$$

where c is understood to be the hypotenuse and a and b are the other sides of the right angle triangle. Usually, two sides of a right triangle are given, and you are required to solve for the third. For example, if $a = 6$ and $b = 8$, then

$$c^2 = 6^2 + 8^2$$
$$c^2 = 36 + 64$$
$$c^2 = 100$$

Taking the square root of the two sides added together will give

$$\sqrt{c^2} = \sqrt{100}$$
$$c = 10$$

Another example could be that one side and the hypotenuse are given. For instance, if $c = 15$ and $a = 9$, then

$$15^2 = 9^2 + b^2$$
$$225 = 81 + b^2$$
$$225 - 81 = b^2$$
$$144 = b^2 \text{ (or } b^2 = 144\text{)}$$

Taking square root of both sides will give

$$b = 12$$

PROBLEM SOLVING WITH TRIANGLES

Here are some questions for you as a teacher to use. You can make up different variations of these questions. The answers are shown in parentheses.

1. If one side of an equilateral triangle is 10 cm, what are the sizes of the other two sides? (10 cm)
2. If two angles of a scalene triangle are 30° and 50°, what is its third angle? (100°)
3. If one angle of an isosceles triangle is 50°, what are the other two angles? (50° and 80° *or* 65° and 65°)
4. If one angle of a right angle triangle is 30°, what are the other two angles? (90° and 60°)
5. If one angle of an equilateral triangle is 60°, what are the other two angles? (60° and 60°)
6. If the adjacent and opposite of a right triangle are 3 cm and 4 cm, what is its hypotenuse? (5 cm)
7. If the hypotenuse and opposite of a right angle triangle are 10 cm and 6 cm, what is its adjacent? (8 cm)

DEFINITIONS OF ANGLES AND INTERSECTING LINES

Congruent angles are two angles that are equal as measured in degrees—in the accompanying example, 55°.

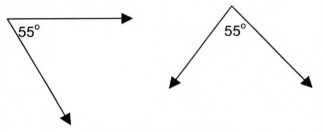

Image 6.2 Congruent Angles

Adjacent angles are two angles in a plane that have a common vertex and a common side. In both pictures, the angles 1 and 2 are adjacent. *Vertex* here is defined as the point where two or more lines intersect.

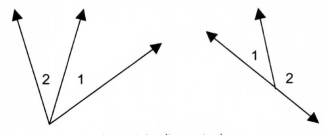

Image 6.3 Adjacent Angles

These concepts are constantly used in physics, engineering, and applied mathematics to prove theorems or to solve problems.

As in the case of triangles, 90° and 180° are two numbers that will occur often when dealing with intersecting lines. Once in a while, 360° becomes important as well in solving problems where the sum of all the angles equals 360°. Some definitions and vocabulary words are important when two lines intersect and form angles.

Complementary angles are two angles whose sum is 90°. Each angle is the *complement* of the other. The angles ∠R and ∠T are complementary because their sum is 90°. Also, angles ∠XYW and ∠WYZ are complementary because their sum, too, is 90°.

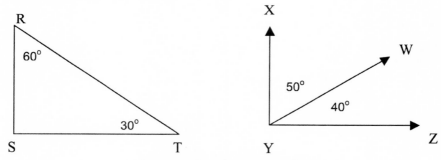
Image 6.4 Complementary Angles

Supplementary angles are angles whose sum is 180°. Each angle is a *supplement* of the other, so the two angles add up to 180°. Angles ∠A and ∠B are supplementary, as are the angles ∠BCE and ∠ECD. In each case, the sum of the two angles is 180°.

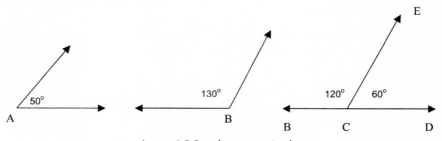
Image 6.5 Supplementary Angles

Vertical angles are formed by the intersection of two lines and are equal. When two lines intersect, they form two pairs of vertical angles. Angles 1 and 3 are vertical angles and are equal. Similarly, angles 2 and 4 are vertical angles and are equal.

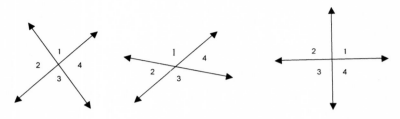

Image 6.6 Vertical Angles

PROBLEM SOLVING WITH ANGLES AND INTERSECTING LINES

Ask students to help you with the following problems.

Example 6.1. Exercises on Adjacent, Vertical, and Complementary Angles

Question	Answer
1. Draw two congruent angles.	
2. Draw two adjacent angles.	
3. If one of the complementary angles is 30°, what is the other angle?	60°
4. If one of the supplementary angles is 70°, what is the other angle?	110°
5. Draw two intersecting lines and identify pair of vertical angles.	
6. If one of the vertical angles in a set is 40°, what is the other vertical angle in this set?	40°
7. Draw two parallel lines and a third line that intersects these lines. Identify vertical angles and supplementary angles.	
8. What are the measures of two congruent and complementary angles?	45°
9. What are the measures of two congruent and supplementary angles?	90°
10. What is the complementary angle that is twice as large as the angle?	60°
11. What is the supplementary angle that is twice as large as the angle?	120°

UNITS OF MEASUREMENT AND THEIR CONVERSION

The three most basic types of measurement are length, mass, and time. Each one is measurable and has a unit. When we say the length of the top of a table is 4 ft, what we are saying is that the measurement of length is 4 using the unit of feet. Of course, there are other types of measurements that we must deal with, but for the time being, what we have will do.

We can measure length in yards, feet, or inches. Similarly, we can find the mass of a body in kilograms or grams. Additionally, we can measure time in days, hours, minutes, or seconds. What we face on a daily basis is the need to convert between these units. To help with this conversion, ask students to provide the following facts because many of them will remember them. The abbreviated form of the unit is indicated in the parentheses.

- 1 yard (yd) = 3 feet (ft)
- 1 ft = 12 inches (in or ")
- 1 kilogram (kg) = 1000 grams (g)
- 1 day = 24 hours (hr)
- 1 hr = 60 minutes (min)
- 1 min = 60 seconds (sec or s)
- 1 kilometer (km) = 1000 meters (m)
- 1 m = 100 centimeters (cm)
- 1 cm = 10 millimeters (mm)

CONVERSION BETWEEN UNITS

Ask students to provide as many answers to the following conversion problems as possible mentally. The answers are shown in parentheses.

1. 5 yd = _____ ft (15 ft)
2. 3 ft = _____ in (36 in)
3. 24 in = _____ ft (2 ft)
4. 12 ft = _____ yd (4 yd)
5. 36 in = _____ yd (1 yd)
6. 5 kg = _____ g (5000 g)
7. 6000 gm = _____ kg (6 kg)
8. 2 days = _____ hr (48 hr)
9. 3 hr = _____ min (180 min)
10. 5 min = _____ sec (300 sec)
11. 3600 sec = _____ hr (1 hr)
12. 120 sec = _____ min (2 min)
13. 120 min = _____ hr (2 hr)

14. 1 km = _____ m (1000 m)
15. 1 m = _____ cm (100 cm)
16. 1 m = _____ mm (1000 mm)
17. 1000 m = _____ km (1 km)
18. 100 cm = _____ m (1 m)
19. 10 mm = _____ cm (1 cm)

MENTAL CHAPTER REVIEW

Ask the following questions of the students, making use of the students' differential knowledge bases. Every student should have a chance to provide an answer or assist in providing an answer. If a student answers incorrectly, ask another student to assist in coming to the correct answer.

Example 6.2. Exercises on the Angles of Three Types of Triangles

Question		Answer
1.	If one side of an equilateral triangle is 12 cm, what are the sizes of the other two sides?	12 cm
2.	If two angles of a scalene triangle are 40° and 60°, what is its third angle?	80°
3.	If one angle of an isosceles triangle is 50°, what are the other two angles?	50°, 80° or 65°, 65°
4.	If one angle of a right angle triangle is 45, what are the other two angles?	90°, 45°
5.	If one angle of an equilateral triangle is 60°, what are the other two angles?	60°, 60°
6.	If the adjacent and opposite of a right angle triangle are 6 cm and 8 cm, what is its hypotenuse?	10 cm
7.	If the hypotenuse and opposite of a right angle triangle are 5 cm and 3 cm, what is its adjacent?	4 cm
8.	Draw two congruent angles.	
9.	Draw two adjacent angles.	
10.	If one of the complementary angles is 40°, what is the other angle?	50°
11.	If one of the supplementary angles is 60°, what is the other angle?	120°

12. Draw two intersecting lines and identify pair of vertical angles.
13. If one of the vertical angles in a set is 45°, what is the other vertical angle? 45°
14. Draw two parallel lines and a third line that intersects these lines. Identify vertical angles and supplementary angles.
15. What are the measures of two congruent and complementary angles? 45°
16. What are the measures of two congruent and supplementary angles? 90°
17. What is the complementary angle that is five times as large as the angle? 75°
18. What is the supplementary angle that is five times as large as the angle? 150°

Ask students to provide as many answers to the following as possible mentally.

Example 6.3. Exercises on Commonly Used Conversions

Convert	Answer
1. 4 yd = _____ ft	12 ft
2. 5 ft = _____ in	60 in
3. 36 in = _____ ft	3 ft
4. 15 ft = _____ yd	5 yd
5. 72 in = _____ yd	2 yd
6. 3 kg = _____ g	3000 g
7. 2000 g = _____ kg	2 kg
8. 3 days = _____ hr	72 hr
9. 4 hr = _____ min	240 min
10. 3 min = _____ sec	180 sec
11. 3600 sec = _____ hr	1 hr

12. 180 sec = _____ min   3 min
13. 180 min = _____ hr   3 hr
14. 2 km = _____ m   2000 m
15. 3 m = _____ cm   300 cm
16. 600 cm = _____ m   6 m
17. 90 mm = _____ cm   9 cm

CHALLENGING PROBLEMS

1. A 15-ft ladder leans against a building. The bottom of the ladder is 9 ft from the wall of the building. How high is the top of the ladder when measured along the wall?
2. The measure of one of a pair of complementary angles is 6° less than twice the measure of the other. Find the measures of both complementary angles.
3. The difference between the measures of two supplementary angles is 40°. What are the two supplementary angles?
4. Find the three angles of a triangle if the second angle is twice the first angle and the third angle is three times the first angle.
5. Find the three angles of a triangle if the second angle is three times the first angle and the third angle is twice the second angle.

MULTIPLICATION TABLES

Write the answers in the spaces provided.

Table 6.1 Multiplication Table up to 13

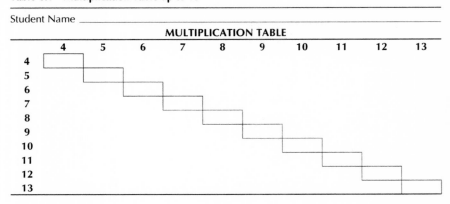

SAMPLE EXIT QUESTIONS

1. $\dfrac{464 \times 569 \times (z - z)}{291 \times 826}$

2. $\dfrac{198 \cdot 5169 \cdot 895}{291 \cdot 826 \cdot (789 - 789)}$

3. Find the three angles of a triangle in which the measure of the second angle is twice the first, and the third angle is equal to the sum of the other two angles.

SAMPLE PRIMING HOMEWORK

Simplify:

1. $212 - 200$
2. $200 - 212$
3. $-30 - 70$
4. $(-8 - 8)$
5. $(-11)(-11)$
6. $\dfrac{-200}{-100}$
7. $11 - 10 + 9 - 8 - 1$
8. $9 + 6 - 2(8 - 10) + (-17)$
9. $\dfrac{-2-3-6}{-9-2}$

POINTS TO REMEMBER AND REVIEW

- The sum of the angles of a triangle is 180°.
- A right triangle has one angle that equals 90°.
- Vertical angles are equal and opposite to each other.
- Complementary angles add up to 90°.
- Supplementary angles add up to 180°.
- In an equilateral triangle, all angles are equal.
- If any students are thinking about a major in engineering, physics, math, or architecture, let them know how important geometry will be to these majors.

7

The Two Great Devils

The Negative Sign and Negative Numbers

In math and science, multiplication with a negative sign (−1, kind of) and/or adding two negative numbers is a common daily occurrence. Yet, we have found that teachers and college professors are often shocked when, for example, their students arrive at an answer of 13 or −1 after adding the two negative numbers − 6 and − 7. Equally bad is multiplying with a negative sign within parentheses, such as −4(−5 − 6). A typical student answer could be a 4 or who knows what else. These examples speak for the "hidden" complexities of these simple operations.

As teachers ourselves, we know that your students must develop a mastery of these procedures. Without this mastery, your students will continue to score low in all sorts of achievement tests, including the SAT and college placement tests. Remember, the achievement tests, by and large, are designed to include those questions where students are most likely to make mistakes. The addition and multiplication of negative numbers are two such examples.

This chapter and the next one are special chapters in this text. The concepts and operations in these two chapters are simple and common, but the contents of these chapters are where students make most of their mistakes. There is, of course, no substitute for practicing over, over, and over again the problems in these chapters and their correct solutions. As a teacher, you can help your students by teaching foolproof methods of dealing with these simple, but at the same time difficult and pesky concepts. In these two chapters, we will emphasize those methods.

ADDITION WITH POSITIVE AND NEGATIVE NUMBERS

Instructions for Adding Positive and Negative Numbers

As a teacher, tell your students to:

1. Treat positive numbers as income coming in to them, money coming into their pockets, for example, allowances from their parents, income from chores, or money earned at a job. In contrast, treat negative numbers as expenditures, money going out of their pockets, say, money spent in buying coffee, a subway fare card, tennis shoes, or food.

    For example, if Bill receives $10 in allowance from his parents, $20 for doing chores in his neighborhood, and $70 over the weekend working in a restaurant, he now has $100 as income in his pocket. We represent this addition operation, as follows: $10 + $20 + $70. Another, more general way of writing this operation would be (+10 + 20 + 70), or even more simply, (10 + 20 + 70). When a sign is missing from the first number in a complex operation, a plus sign is assumed. Show these operations on the chalkboard or a flip chart.
2. Group all the positive terms together. Group all the negative terms together.
3. Add all positive terms for one sum; then add all negative terms to get their sum.
4. When students are adding two or more negative terms, *they should add all of them as if they are all positive, but remember to insert the negative sign afterward.*
5. Perform the indicated operation. When you are dealing with both a positive and a negative term together, *subtract* the smaller term from the bigger one, and *place the sign of the bigger term for the answer.*

Illustrating Addition of Positive and Negative Numbers

Example 7.1. Addition and Subtraction of Positive and Negative Integers

Mathematical Operation	Explanation
10 − 20 − 30 − 50 + 30 + 40 + 20 − 60	Adding multiple terms
= 10 + 30 + 40 + 20 − 20 − 30 − 50 − 60	Collecting positive and negative terms
=100 − 160	Adding all positive terms gives us 100.
= − 60	Add all negative terms together, and then add all these terms first and place a negative sign in front of it, which gives -160

Mathematical Operation	Explanation
	When dealing with a subtraction, subtract the smaller term from the larger one, and then place the sign of the bigger term for the answer. In this case the answer is −60.

Now I want you to try some problems that involve addition of positive and negative terms. These problems can either be done orally or be presented as a handout. We suggest orally, involving each of your students—calling each student by name and using the principle of differential knowledge bases with each of them.

Simplify:

1. −8 − 6 =
2. −11 − 6 =
3. −9 − 2 =
4. −14 − 12 =
5. −8 − 2 =
6. −12 − 5 =
7. −11 − 4 =
8. −9 − 1 =
9. −8 − 2 =
10. −10 − 3 =
11. 5 − 2 − 2 =
12. 10 − 4 − 6 =
13. 7 − 2 − 5 =
14. 8 − 2 − 1 =
15. 9 − 1 − 7 =
16. 5 + 1 − 6 =
17. 6 + 2 − 9 =
18. 5 − 3 + 2 − 6 =
19. 6 − 2 − 2 =
20. 12 − 4 − 6 =
21. 9 − 2 − 5 − 2 =
22. 8 − 2 − 1 − 1 =
23. 9 − 1 − 7 − 2 =
24. −2 − 3 + 7 =
25. −5 − 4 + 8 =
26. −3 + 3 − 5 + 5 =
27. −6 + 2 + 5 − 2 =
28. −7 + 2 −2 + 7 =
29. 5 − 7 + 2 − 3 =
30. 6 + 1 − 4 − 4 =
31. −2 − 3 + 5 =
32. −5 − 4 + 9 =
33. −3 + 3 − 5 + 5 =
34. −6 + 2 + 7 − 1 =
35. −7 + 2 − 1 + 7 =

MULTIPLICATION INVOLVING NEGATIVE NUMBERS

The Rules Involving Negative Numbers

Example 7.2. Multiplying Positive and Negative Integers

Number Type	Math Operator	Number Type	Result	Example
Positive number	×	Positive number	Positive	6 x 3 = 18
Positive number	×	Negative number	Negative	−6 x 3 = −18
Negative number	×	Negative number	Positive	−6 x -3 = 18

Multiplication Involving Negative Numbers and Parentheses

In multiplying numbers that are outside and inside the parentheses, simplify the terms in the parentheses first, if possible, before proceeding with multiplication. Remember, when there is no number immediately placed before the open parentheses, tell the students they can assume that there is a 1 in that position. Be extremely careful with this type of multiplication because this is where many of the students will make careless mistakes, and it is one of the main reasons for scoring low on achievement tests.

Simplifying Terms Inside and Outside Parentheses

Let us take two examples first.

Example 7.3. Simplifying Terms in Parentheses

	Example	Explanation
1.	−12 + (17 − 5) = −12 + 12 = 0	17 and 5 are within the parentheses. Simplify within the parentheses, i.e., (17-5) = 12 Perform the final math operation of subtraction.
2.	− 8 − (15 − 7) = − 8 − (8) = − 8 − 8 = −16	15 and 7 are within the parentheses. Simplify within the parentheses, i.e., (15-7) = 8 Perform the final math operation of adding two negative numbers.

Simplifying after Solving within the Parentheses

You, as the teacher, should either write these problems on the chalkboard or prepare a handout for students before engaging students in their learning.

1. $-12 + (17 - 5) =$
2. $-15 + (22 - 7) =$
3. $-8 - (15 - 7) =$
4. $-10 - (22 - 20) =$
5. $-5 + (17 - 12) + 10 =$
6. $(-12 + 17) - 5 =$
7. $-15 + (20 - 5) =$
8. $-11 - (15 - 4) =$
9. $-10 - (22 - 12) =$
10. $-5 - (17 - 12) - 10 =$
11. $-24 + 2(17 - 5) =$
12. $-15 + 2(22 - 7) =$
13. $-24 - 3(15 - 7) =$
14. $-10 - 2(22 - 20) =$
15. $-5 + 3(17 - 12) + 10 =$
16. $4(-12 + 17) - 5 =$
17. $-15 + 2(20 - 5) =$
18. $-11 - 3(15 - 4) =$
19. $-10 - 6(22 - 12) =$
20. $-5 - 8(17 - 12) - 10 =$
21. $(-2)(-5) =$
22. $(-5)(3) =$
23. $12 + -4 + 2 - -6 =$
24. $8 + -4 + 2 - -6 + -12 =$
25. $-9 + -4 + 2 - -6 + -7 + 12 =$
26. $-5 + (2 - 7) - (-2 - 4) =$
27. $-4 + 2(2 - 7) - (6 - 4) =$

MENTAL CHAPTER REVIEW

Students should do some of these problems orally. For other problems, those that are longer and involve more than two terms, students should show their work on the chalkboard. You can also use a handout for these problems. Remember, in addition to keeping your students awake, an interactive approach allows you to quickly deal with errors.

1. $-9 - 6 =$
2. $-12 - 6 =$
3. $-10 - 2 =$
4. $-15 - 12 =$
5. $-10 - 2 =$
6. $-13 - 5 =$
7. $-14 - 4 =$
8. $-11 - 1 =$
9. $-9 - 2 =$
10. $-14 - 3 =$
11. $5 - 3 - 4 =$
12. $10 - 5 - 7 =$
13. $7 - 3 - 6 =$
14. $8 - 2 - 3 =$

15. 9 − 2 − 7 =
16. 5 + 1 − 6 =
17. 6 + 4 − 9 =
18. 5 − 3 + 4 − 6 =
19. 7 − 2 − 2 =
20. 13 − 4 − 6 =
21. 11 − 2 − 5 − 2 =
22. 10 − 2 − 1 − 1 =
23. 8 − 1 − 7 − 2 =
24. −4 − 3 + 7 =
25. −6 − 4 + 8 =

26. −4 + 3 − 5 + 5 =
27. −8 + 2 + 5 − 2 =
28. −9 + 2 − 2 + 7 =
29. 7 − 7 + 2 − 3 =
30. 9 + 1 − 4 − 4 =
31. −4 − 3 + 5 =
32. −6 − 4 + 9 =
33. −6 + 3 − 5 + 5 =
34. −7 + 2 + 7 − 1 =
35. −11 + 2 − 1 + 7 =

Simplify after solving within parentheses (Caution: when you see a term like − −8, what it really means is that the number 8 is negative and is also being multiplied by a negative 1. So, the answer is +8).

36. −14 + (17 − 5) =
37. −16 + (22 − 7) =
38. −9 − (15 − 7) =
39. −8 − (22 − 20) =
40. −7 + (17 − 12) + 11 =
41. (−10 + 17) − 5 =
42. −16 + (20 − 5) =
43. −13 − (15 − 4) =
44. −11 − (22 − 12) =
45. −6 − (17 − 12) − 10 =
46. −25 + 2(17 − 5) =
47. −16 + 2(22 − 7) =
48. −26 − 3(15 − 7) =
49. −12 − 2(22 − 20) =

50. −4 + 3(17 − 12) + 14 =
51. 4(−11 + 17) − 5 =
52. −16 + 2(20 − 15) =
53. −11 − 3(15 − 6) =
54. −12 − 6(22 − 14) =
55. −7 − 8(17 − 13) − 10 =
56. (−6)(−7) =
57. (−9)(8) =
58. 13 + −4 + 2 − −6 =
59. 11 + −4 + 2 − −6 + −12 =
60. −11 + −4 + 5 − −6 + −7 + 12 =
61. −5 + (2 − 8) − (−3 − 4) =
62. −4 + 2(3 − 7) − (5 − 4) =

CHALLENGING PROBLEMS

Do these problems in class, engaging students by calling on them by name and using their differential knowledge bases. In this way, you will know right away who needs help and whether you need to review the concepts involved for some or all of your students.

Simplify:

1. $(-1)(-1)(-1) =$
2. $(-1)(-1)(-1)(-1) =$
3. $(-1)^5 =$
4. $(-1)^6 =$
5. $(-1)^{101} =$
6. $(-1)^{100} =$
7. $(-1)(268) =$
8. $(-1)^5(234) =$
9. $(-1)^{50}(56) =$
10. $2048 \times 1 =$
11. $4096 / 1 =$
12. $9123 - 9123 =$

MULTIPLICATION TABLES

Table 7.1 Multiplication table up to 14

Write the answers in the spaces provided.

Student Name :_____

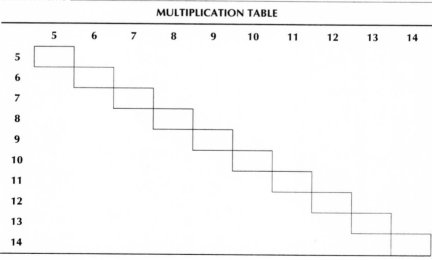

SAMPLE EXIT QUESTIONS

Simplify:

1. $-9 + (7 - 8) - (-5 - 6) =$
2. $(-1)(-1) =$
3. $(-1)^3 =$
4. $(-1)^4 =$
5. $(-1)^{101} =$
6. $(-1)^{99} =$
7. $[3 + 11^0 + (-1)^6 - (-1)^9] =$
8. 310 cm = _____ m

SAMPLE PRIMING HOMEWORK

Simplify:

1. $2\frac{3}{4} + \frac{1}{4} =$
2. $2\frac{3}{4} - \frac{3}{4} =$
3. $2\frac{3}{4} \times \frac{3}{11} =$
4. $2\frac{3}{4} \div \frac{15}{4} =$
5. $-31 + 3(17 - 7) + (y - z)^0 =$
6. $[3 + (4 - 6) * 10^0 - x^0 - \frac{5*6*3}{6*15}] =$

POINTS TO REMEMBER AND REVIEW

- Group all of the positive terms together and all the negative terms together.
- Add all positive terms for one sum; then add all negative terms to get their sum.
- When adding two or more negative terms, add them as if they are all positive, but remember the negative sign. After completing the adding operation, put a negative sign in front of this sum.
- When you are dealing with both positive and negative terms together, subtract the smaller term from the bigger one (magnitude-wise) and place the sign of the bigger term for the answer.
- An even power of -1, e.g., $(-1)^{16}$, gives a positive answer, whereas an odd power of -1, e.g., $(-1)^{17}$, will give a negative answer.

8
Change for Currency Notes

Simple Fractions

Let your students know that, like statistics, fractions are all around us. For example, 50 cents is 1/2 of a dollar; 25 cents is 1/4 of a dollar; a dime is 1/10 of a dollar. All these examples are really fractions. As another pair of examples, when the students go to a grocery store, they deal with half (1/2) gallons and quarts—that is, quarter (1/4) gallons. Also, when dealing with measurements in inches, we frequently come across fractions like 1/2, 1/4, 1/8, or 1/16. A number like 7/16 is also a fractional number, with 7 being called the *numerator* of the fraction and 16 being the *denominator*.

In addition, there are two definitions your students need to know about fractions. A *proper fraction* is one whose numerator is less than its denominator (e.g., $\frac{3}{4}$ or $\frac{5}{8}$). An *improper fraction* is the opposite, where the numerator is equal to or greater than the denominator (e.g., $\frac{4}{3}$ or $\frac{7}{7}$). They may not see examples of improper fractions here, but warn the students they will see improper fractions in later chapters.

Recognition of a fraction, even before doing an arithmetic operation, is important. For example, 16/4 is simply 4, and 12/24 is 1/2. Please keep in mind that for ease of writing, simplicity, and space saving, a fraction like $\frac{5}{8}$ can be written as 5/8. These two methods of writing are equivalent.

Commonly Used Operations with Fractions

Addition with Fractions   *Practical Example*

$\frac{1}{2} + \frac{1}{2} = 1$ 50 cents + 50 cents = one dollar

$\frac{1}{4} + \frac{1}{4} = \frac{1}{2}$ 25 cents + 25 cents = 50 cents

$\frac{1}{2} + \frac{1}{4} = \frac{2}{4} + \frac{1}{4} = \frac{3}{4}$ 50 cents + 25 cents = 75 cents = 3/4 of a dollar

$\frac{1}{4} + \frac{1}{4} + \frac{1}{4} + \frac{1}{4} = 1$ 4 quarts of milk = 1 gallon of milk

$\frac{1}{4} + \frac{1}{4} = \frac{1}{2}$ 1/4 inch + 1/4 inch = 1/2 inch

$\frac{1}{8} + \frac{1}{8} + \frac{1}{8} + \frac{1}{8} = \frac{1}{2}$ This is equal to 4/8 = 1/2 inch

$\frac{1}{2}$ divided by $2 = \frac{1}{4}$ Cutting half of a cake into two (1/4 of a cake)

$\frac{1}{2}$ multiplied by $2 = 1$ Adding 1/2 to itself = 1

$1 + \frac{1}{2} = 1\frac{1}{2}$ 1 dollar and 50 cents = 1½ dollars (mixed numeral)

Mental Addition and Subtraction of Fractions

If you think there is a need for further illustrations, you should demonstrate several examples before asking students to do the following list mentally. However, you need to emphasize the fact that *when denominators are the same, you simply add the numerators.* Also, ask students to recognize *quickly* that 4/4 is 1 and 16/4 is 4.

1. $\frac{1}{4} + \frac{1}{4} =$

2. $\frac{1}{2} + \frac{1}{4} =$

3. $\frac{1}{4} + \frac{1}{2} =$

4. $\frac{1}{4} + \frac{1}{4} + 1 =$

5. $\frac{1}{4} + \frac{3}{4} =$

6. $\frac{1}{4} + \frac{1}{1} =$

7. $\frac{1}{2} + \frac{1}{1} =$

8. $\frac{1}{4} + \frac{1}{4} + \frac{1}{4} =$

9. $\frac{1}{4} + \frac{1}{4} + \frac{1}{4} + \frac{1}{4} =$

10. $\frac{1}{4} + \frac{1}{4} + \frac{1}{2} =$

11. $1 + \frac{1}{4} - \frac{1}{4} =$

12. $2 + \frac{1}{2} + \frac{1}{2} - \frac{1}{2} - \frac{1}{2} =$

13. $\frac{1}{2} + \frac{1}{2} + 3 - 1 =$

14. $\frac{1}{2} + \frac{1}{2} + 4 =$

15. $\frac{1}{4} - \frac{1}{4} + \frac{3}{4} + \frac{1}{4} =$

16. $\frac{1}{4} + \frac{1}{4} + \frac{3}{6} =$

17. $\frac{1}{4} + \frac{1}{4} + \frac{4}{8} =$

18. $\frac{1}{4} + \frac{3}{4} + 2 - 1 =$

19. $\frac{1}{4} + \frac{3}{4} + 2 =$

20. $\frac{1}{8} + \frac{1}{8} =$

21. $\frac{3}{8} + \frac{1}{8} - \frac{1}{8} =$

22. $\frac{1}{8} + \frac{1}{8} + \frac{1}{8} + \frac{1}{8} =$

23. $\frac{1}{2} + \frac{1}{2} + \frac{2}{2} - \frac{2}{2} + 2 =$

24. $\frac{1}{8} + \frac{1}{8} + \frac{1}{8} + \frac{1}{8} + \frac{4}{8} =$

Multiplication of Simple Fractions (Do mentally)

If there is a need, you should demonstrate several examples before asking students to do the following list mentally. However, during multiplication of fractions, emphasize using the technique of canceling the same factors in the numerator and denominator and simplifying. Once the fractions are simplified, multiply all the numerators and all the denominators.

Emphasize to the students that they should *simplify before multiplying* when dealing with fractions. For example, a fraction like: $\frac{200}{400} \times \frac{1000}{500}$ looks quite complex, but the answer is 1. This is because simplification or cancellation of zeros in the numerator and denominator will give you 2/2, which is equal to 1. You should either prepare a handout or write these fractions on the chalkboard or a combination of the two before you ask students to simplify them mentally.

1. $\frac{1}{2} * \frac{1}{4} =$

2. $\frac{1}{4} * \frac{1}{8} =$

3. $\frac{1}{4} * \frac{1}{4} =$

4. $\frac{1}{4} * \frac{4}{4} =$

5. $\frac{1}{4} * \frac{1}{1} =$

6. $\frac{1}{2} * \frac{1}{1} =$

7. $\dfrac{1}{4} * \dfrac{1}{4} * \dfrac{1}{4} =$

8. $\dfrac{1}{4} * \dfrac{1}{4} * \dfrac{1}{2} =$

9. $\dfrac{1}{4} * \dfrac{1}{2} * \dfrac{1}{2} =$

10. $\dfrac{3}{4} * \dfrac{1}{4} =$

11. $\dfrac{1}{2} * \dfrac{1}{2} * \dfrac{1}{2} =$

12. $\dfrac{4}{2} * \dfrac{1}{2} =$

13. $\dfrac{4}{2} * \dfrac{1}{4} =$

14. $\dfrac{1}{4} * \dfrac{1}{2} * \dfrac{8}{1} =$

15. $\dfrac{1}{4} * \dfrac{1}{4} * \dfrac{16}{1} =$

16. $\dfrac{1}{4} * \dfrac{1}{4} * \dfrac{4}{8} =$

17. $\dfrac{1}{4} * \dfrac{3}{4} * \dfrac{16}{3} =$

18. $\dfrac{1}{4} * \dfrac{3}{4} * \dfrac{32}{3} =$

19. $\dfrac{1}{8} * \dfrac{1}{8} * \dfrac{16}{1} =$

20. $\dfrac{1}{8} * \dfrac{1}{8} * \dfrac{32}{1} =$

21. $\dfrac{1}{8} * \dfrac{1}{8} * \dfrac{4}{1} * \dfrac{16}{1} =$

22. $\dfrac{1}{4} * \dfrac{1}{2} * \dfrac{1}{8} * \dfrac{8}{1} * \dfrac{16}{1} =$

Division of Simple Fractions (Do mentally)

If necessary, you should demonstrate several examples before asking students to complete the following list mentally. Also point out that students *first* need to change the division sign to multiplication and invert the term (swap numerator and denominator—as the reciprocal) in the fraction term that *follows* the division sign. You, as the teacher, might decide to illustrate this process if the students seem to have a problem with it. *Second*, simplify the terms if possible, before proceeding to multiplication, just as was done in the previous section.

1. $\dfrac{1}{2} \div 2 =$

2. $\dfrac{1}{4} \div 2 =$

3. $\dfrac{1}{8} \div \dfrac{1}{8} =$

4. $\dfrac{1}{2} \div \dfrac{1}{2} =$

5. $\dfrac{1}{4} \div \dfrac{1}{4} =$

6. $\dfrac{1}{64} \div \dfrac{1}{128} =$

7. $\dfrac{1}{24} \div \dfrac{1}{96} =$

8. $\dfrac{5}{4} \div \dfrac{5}{4} =$

9. $\dfrac{1}{4} \div \dfrac{1}{8} =$

10. $\dfrac{1}{8} \div \dfrac{1}{16} =$

11. $\dfrac{1}{16} \div 0 =$

12. $\dfrac{1}{2} \div 1 =$

13. $\dfrac{1}{2} \div \dfrac{2}{4} =$

14. $\dfrac{1}{2} \div \dfrac{4}{8} =$

15. $\dfrac{1}{8} \div \dfrac{8}{64} =$

16. $\dfrac{1}{4} \div \dfrac{4}{16} =$

Word Problems

Stress the key words and their relationship to arithmetic operations in the following word problems.

1. How much will it cost you to buy four candy bars if each candy bar costs a quarter?
2. You have two wooden blocks. The width of each block is 1/4 of an inch. If you put them together, what will be the total width?
3. Your parents gave you $20.00 for your weekly allowance. You went to the mall with your friends and went on a mini spending spree. You bought a special drink for $3.25, chocolate candy for $1.25, a paperback book for $8.50, and French fries for $1.00. How much do you still have at the end?
4. You and your friend were supposed to attend a birthday party for another friend. The two of you arrived very late for it. By the time you got there, the guests had consumed 3/4 of the cake. The rest of the cake was divided equally between the two of you. What fraction of the total cake did you actually get?
5. You and two friends were doing an experiment in measurement. You were supposed to find the volume (length × width × height) of a three-dimensional block of wood. You measured the length to be 2 inches; the two other measurements of width and height were 1 inch and 1/2 inch. What is the volume of the block in cubic inches?

MENTAL CHAPTER REVIEW

Involve all of the students in providing answers to the following mental review. Make certain to use the pedagogical technique of differential knowledge bases and individual student names when selecting students to answer these questions. Do not allow students to call out, or you will be down to the three smug swifties in the front seats.

Perform the indicated operations.

1. $\frac{1}{2} + \frac{1}{2} =$
2. $\frac{1}{4} + \frac{1}{4} =$
3. $\frac{1}{3} + \frac{1}{3} + \frac{1}{3} =$
4. $\frac{1}{4} + \frac{3}{4} =$
5. $\frac{1}{2} + \frac{3}{2} =$
6. $\frac{1}{4} + \frac{7}{4} =$
7. $\frac{1}{2} + \frac{1}{4} + \frac{1}{4} =$
8. $\frac{3}{4} + \frac{5}{4} =$
9. $\frac{4}{3} + \frac{2}{3} =$
10. $\frac{6}{2} + \frac{6}{2} =$
11. $\frac{12}{3} + 6 =$
12. $\frac{1}{2} - \frac{1}{4} =$
13. $\frac{1}{4} - \frac{1}{8} =$
14. $\frac{1}{3} - \frac{1}{6} =$

15. $\frac{1}{5} - \frac{1}{10} =$
16. $\frac{3}{4} - \frac{1}{4} =$
17. $\frac{5}{4} - \frac{1}{4} =$
18. $\frac{6}{5} - \frac{1}{5} =$
19. $\frac{8}{3} - \frac{2}{3} =$
20. $\frac{9}{12} - \frac{3}{12} =$
21. $\frac{11}{36} - \frac{11}{36} =$
22. $\frac{9}{7} - \frac{2}{7} =$
23. $\frac{1}{2} \cdot 2 =$
24. $\frac{1}{4} \cdot 4 =$
25. $\frac{1}{5} \cdot 10 =$
26. $\frac{1}{2} \cdot 4 =$
27. $\frac{1}{4} \cdot 8 =$
28. $\frac{1}{5} \cdot 15 =$

29. $\frac{1}{6} \cdot 18 =$
30. $\frac{1}{2} \cdot \frac{8}{2} =$
31. $\frac{8}{4} \cdot \frac{25}{5} =$
32. $\frac{8}{32} \cdot \frac{4}{1} =$
33. $\frac{1}{2} \div 2 =$
34. $\frac{1}{4} \div 2 =$
35. $\frac{1}{5} \div \frac{1}{5} =$
36. $\frac{1}{2} \div \frac{1}{2} =$
37. $\frac{5}{2} \div 5 =$
38. $\frac{2}{5} \div \frac{2}{5} =$
39. $\frac{5}{3} \div \frac{15}{3} =$
40. $\frac{4}{8} \div \frac{1}{2} =$
41. $\frac{6}{12} \div 2 =$
42. $\frac{8}{2} \div \frac{4}{1} =$

SAMPLE EXIT QUESTIONS

1. $1 - \frac{1}{2} - \frac{1}{2} =$
2. $2\frac{3}{4} - \frac{3}{4} =$
3. $3\frac{1}{2} - 2\frac{1}{4} =$
4. $\sqrt{196} =$
5. $\frac{1}{2} + \frac{1}{2} + \frac{3}{4} + \frac{3}{4} - 2\frac{1}{2} =$
6. $(-1)^{97}(p) =$

SAMPLE PRIMING HOMEWORK

Write in powers of 10:

1. $100 =$
2. $1{,}000{,}000 =$

Simplify:

3. $(10^3)(10^5) =$
4. $(10^4)(10^4) =$
5. $(10^4)^2 =$
6. $\dfrac{10^3}{10^3} =$
7. $\dfrac{10^3}{10^{-6}} =$

Write in scientific notation:

8. $345{,}792.34$
9. 0.000689

POINTS TO REMEMBER AND REVIEW

- Just know some of the simple fractions by heart. For example, 1/2 is 50 cents in currency, 0.5 in decimal notation, or 50% in percent notation.
- When you see a division sign, change that sign to a multiplication sign and take the reciprocal of the term that follows the division sign.
- Simplify in the numerator and denominator before multiplying or dividing.
- In complex terms, insert 1 as the denominator for a term that shows only the numerator, e.g., $\dfrac{1}{2} \div 2 = \dfrac{1}{2} \div \dfrac{2}{1}$ (both sides are equivalent)
- Try to recognize patterns when you encounter something difficult looking, such as $\dfrac{100*10}{1000} - \dfrac{22*33*2}{121*12}$ (the answer for this expression is zero—the first term simplifies to 1, and the second term simplifies to 1 as well).

9

Exponents

Powers of 10

In math, science, and engineering, we come across some numbers that are either very large—such as 5,980,000,000,000,000,000,000,000 kilograms for the mass of the Earth—or very tiny, like 0.00000000000000000000000000167 kilograms for the mass of a proton. As the teacher, writing these numbers on the board for the students to see and understand that these numbers written out with all their respective zeros takes too long and occupies too much space. Also, these numbers are much harder to read if written in such a long format; even worse, most of us are going to make mistakes when writing numbers in this format.

So, scientists and engineers have devised a more accurate and shorter way of writing these gargantuan and infinitesimal numerals, using a nifty technique: the powers of 10. Even the U.S. federal budget ($1,800,000,000,000 in 2004) has grown to the size where it almost needs to be written in scientific notation using powers of 10, e.g., 1.8×10^{12} dollars. A spreadsheet may write this number as 1.8E+12 in scientific notation, where "E+12" is the equivalent of "$\times 10^{12}$"; E, in this case, means the base 10.

This chapter introduces the powers of 10. Then it explains arithmetic operations involving such powers. However, for all of us to be together, we want to be sure of your students' understanding of what we have been doing before we take up powers of 10.

MENTAL REVIEW OF PREVIOUS KNOWLEDGE

Review the information in this first section of the chapter with your students' help during the first 15–20 minutes of the class period. This review of prior chapters does not mean that you have to reteach these concepts, unless absolutely essential, but if a substantial number of students are having some difficulties with any particular topic under review, *stop right there and reteach the material.* Do not move on until everybody is with you—easily. Once again, involve every student in answering these questions based on the pedagogical principles of their differential knowledge bases and equal participation.

Review of Basic Numbers and Arithmetic Operations

All of your students should demonstrate that they are able to provide answers mentally to all these problems before you move to the next section in this chapter. Also, being able to handle all of these items boosts the students' self-confidence in doing math. This review section can be done using oral questions and answers, a chalkboard where appropriate, or pencil and paper where necessary. Vary the ways of producing the answers.

1. Write a word name for 413,008.
2. Write a word name for 5,617,018.
3. Write a number for five million, two hundred ninety-one thousand, three hundred forty-one.
4. Write a number for fifty-six million, three hundred thousand, three hundred forty-one.

Simplify:

5. $550 + 550 =$
6. $998 + 1002 =$
7. $2048 - 1008 =$
8. $4096 - 2090 =$
9. $13 \times 13 =$
10. $11 \times 11 =$
11. $72 / 8 =$
12. $200 / 20 =$
13. $8 * 8 =$
14. $7 * 7 =$
15. $6^2 =$
16. $9^2 =$
17. $(81)^{1/2} =$
18. $(64)^{1/2} =$
19. $13 / 0 =$
20. $234 / 0 =$
21. $\frac{1}{2} \div 2 =$
22. $\frac{1}{2} + \frac{5}{2} =$
23. $\frac{10}{2} + \frac{20}{2} =$
24. $\frac{28}{4} + \frac{40}{8} =$
25. $\frac{7}{2} \div \frac{7}{2} =$
26. $\frac{8}{2} \times \frac{6}{3} =$
27. $\frac{48}{12} + \frac{66}{11} =$
28. $\frac{400}{4} - \frac{200}{4} =$
29. $\frac{1}{4} \div 4 =$

Review of Facts about Our Sometimes Pesky Friends, 1 and 0

1. $32 \times 1 =$
2. $0 + 512 =$
3. $16/1 =$
4. $256 + 0 =$
5. $128/1 =$
6. $1 \times 64 =$
7. $0 * 16 =$
8. $0/9 =$
9. $11 * 0 =$
10. $0/45 =$
11. $9/0 =$
12. $10/0 =$
13. $100/0 =$
14. $0/156 =$
15. $34/0 =$

Zero as Exponent

Here is our old pal, the zero, back again with some technical tricks. Fact: Any number or variable x, raised to the power of 0, gives a result of 1. For example:

$$8^0 = 1$$
$$10^0 = 1$$
$$x^0 = 1$$
$$N^0 = 1$$

Your students should mentally provide the answers for the following problems.

1. $10^0 =$
2. $100^0 =$
3. $50^0 =$
4. $212^0 =$
5. $X^0 =$
6. $(2X)^0 =$
7. $(X/2)^0 =$
8. $(5X)^0 =$
9. $N^0 =$
10. $(9N)^0 =$
11. $(3Y)^0 =$
12. $(6T)^0 =$
13. $F^0 =$
14. $(2U)^0 =$
15. $(5/C)^0 =$

POWERS OF 10

Positive and Negative Exponents to Be Overlearned

Be sure that students are familiar with some standard powers of 10 before they proceed to the next sections in this chapter. Introduce this section by saying, "Now we are going to deal with a notation type called the 'powers of 10,' in which we will learn about the gigantic range of numbers in our lives. Some will be huge, and some will be microscopic. Learning to use the powers of 10 will save you all kinds of headaches, and they are very important in the sciences, such as physics, astronomy, biology, engineering, and statistics." Introduce the powers of 10 in the order of familiarity (from more familiar to less familiar).

Table 9.1 Standard powers of 10

Number	Name	Power of 10 Notation
1,000,000,000,000	Trillion	10^{12}
1,000,000,000	Billion (also called giga and denoted by G)	10^9
1,000,000	Million (also called mega, and denoted by M)	10^6
1000	Thousand (also called Kilo and denoted by K)	10^3
100	One hundred	10^2
10	Ten	10^1 or 10
1	One	10^0

Exponents

Number	Name	Power of 10 Notation
$\frac{1}{10} = 0.1$	One-tenth	10^{-1}
$\frac{1}{100} = 0.01$	One-hundredth	10^{-2}
$\frac{1}{1000} = 0.001$	Thousandth (also called milli)	10^{-3}
$\frac{1}{10000} = 0.0001$	Millionth (also called micro)	10^{-6}
$\frac{1}{1000000000} = 0.000000001$	Billionth	10^{-9}
$\frac{1}{1000000000000} = 0.000000000001$	Trillionth (also called Pico)	10^{-12}

Exercise Set 9.1

Verbalize and write the following in power-of-10 notation.

Engage each student to say the answers for the questions that follow—loud and clear. Some of the problems in this exercise set involve the concept of scientific notation. Tell the students to ignore this concept for the time being if they are not familiar with it. We will introduce it later. The concept is included here to test the knowledge bases of students for powers of 10.

1. 100
2. 10,000
3. 1,000,000
4. 100,000,000
5. 10,000,000,000
6. 100,000,000,000
7. 10
8. 1000
9. 100,000
10. 10,000,000
11. 1,000,000,000
12. 100,000,000,000
13. 0.1
14. 0.001
15. 0.00001
16. 0.0000001
17. 0.000000001
18. 0.01
19. 0.0001
20. 0.000001
21. 0.00000001
22. 0.0000000001
23. 50,000,000
24. 30,000
25. 900,000,000
26. 60,000,000,000
27. 0.3
28. 0.005
29. 0.00006
30. 0.0000008

MATHEMATICAL OPERATIONS INVOLVING POWERS OF 10

When dealing with the powers of 10 (e.g., 10^5), 10 is called the *base* and 5, in this instance, is the *power of 10*. Knowing and remembering this terminology is important when dealing with such powers. Remind students that the *power* of the base is *the number of times the base is multiplied with itself*. For example, 10^3 means that 10, the base, is multiplied by itself three times, giving an answer of $10 \times 10 \times 10 = 1000$.

Multiplication

In multiplication involving powers of the same bases—in this case, 10—you add the powers of the base to obtain a final answer. If you see only the base 10, without any power, it is assumed that it has a power of 1; 10 is the same as 10^1. Involve students in solving the following problems; the answers are shown for each problem.

1. $10 \times 10 = 10^1 \times 10^1 = 10^2$
2. $10 \times 10^2 = 10^3$
3. $10^2 \times 10^3 = 10^5$
4. $10^6 \times 10^6 = 10^{12}$
5. $10^4 \times 10^9 = 10^{13}$
6. $10^{-1} \times 10^{-1} = 10^{-2}$
7. $10^{-3} \times 10^{-3} = 10^{-6}$
8. $10^{-4} \times 10^{-8} = 10^{-12}$
9. $10^4 \times 10^{-3} = 10^1$
10. $10^8 \times 10^{-10} = 10^{-2}$
11. $10^4 \times 10^{-4} = 10^0 = 1$
12. $10^7 \times 10^{-7} = 10^0 = 1$

Division

In division involving powers *when base is the same*—or more specifically in this case, when the base is 10—*subtract the powers* when finding the final answer. For example:

$$\frac{10^6}{10^2} = 10^{6-2} = 10^4.$$

Another way to solve this problem is:

$$\frac{10^6}{10^2} = 10^6 \times 10^{-2} = 10^{6+(-2)} = 10^4.$$

The principle that we have used here is that *when you bring quantities involving powers from denominator to numerator and vice versa, the power changes sign*. Feel free to use other examples to illustrate this concept further, if needed.

Use the following examples to get answers from students. Write some of these problems on the chalkboard to clarify the division and the powers. Do not forget our old pal, zero, because $10^0 = 1$.

1. $10 / 10 = 10^1 \times 10^{-1} = 10^{1-1} = 10^0 = 1$
2. $10^2 / 10^1 = 10^{2-1} = 10^1$
3. $10^6 / 10^6 = 10^{6-6} = 10^0 = 1$
4. $10^4 / 10^9 = 10^{4-9} = 10^{-5}$
5. $10^{-3} / 10^{-3} = 10^{-3+3} = 10^0 = 1$
6. $10^{-4} / 10^{-8} = 10^{-4+8} = 10^4$
7. $10^8 / 10^{-10} = 10^{8+10} = 10^{18}$
8. $10^4 / 10^{-4} = 10^{4+4} = 10^8$
9. $10^2 / 10^3 = 10^{2-3} = 10^{-1}$
10. $10^{-1} / 10^{-1} = 10^{-1+1} = 10^0 = 1$

Exercise Set 9.2

Simplify the indicated operations and write the answer into powers of 10. For this exercise, engage all of the students in providing answers to the following questions.

1. $10 \times 1 =$
2. $10^1 \times 10 =$
3. $10 \times 10^1 =$
4. $10^3 \times 10^2 =$
5. $10^4 \times 10^4 =$
6. $10^{-1} \times 10^{-1} =$
7. $10^{-2} \times 10^{-4} =$
8. $10^{-3} \times 10^{-11} =$
9. $10^5 \times 10^{-3} =$
10. $10^8 \times 10^{-13} =$
11. $10^1 \times 10^{-4} =$
12. $10^9 \times 10^{-9} =$
13. $10 / 10 =$
14. $10^3 / 10^2 =$
15. $10^5 / 10^8 =$
16. $10^{11} / 10^6 =$
17. $10^{14} / 10^8 =$

18. $10^{14} / 10^{12} =$

19. $10^{-4} / 10^{-4} =$

20. $10^{-4} / 10^{-12} =$

21. $10^{19} / 10^{11} =$

22. $10^{10} / 10^{-10} =$

23. $10^{20} / 10^{-20} =$

24. $10^{19} / 10^{5} =$

25. $10^{2} / 10^{3} =$

26. $10^{-1} / 10^{-1} =$

27. $10^{5} / 10^{5} =$

28. $10^{15} / 10^{5} =$

29. $10^{16} / 10^{-4} =$

30. $10^{20} / 10^{-10} =$

Addition and Subtraction

Addition or subtraction involving powers of 10 can be quite tricky. When students see an addition or subtraction that involves powers of 10, make certain they look carefully at the powers. If the powers are not the same, they must first *make the powers the same*. Otherwise, the answer will be incorrect. Once assured that the powers are the same, only then can addition or subtraction of the coefficients proceed. In an expression like 5×10^2, 5 is the *coefficient*.

Some examples of addition and subtraction that involve powers of 10 are:

Example 9.1. Adding and Subtracting Terms Involving Powers of 10

Example	Explanation
1. $3 \times 10^3 + 4 \times 10^3$ $= 7 \times 10^3$	In this case, the powers of 10 are the same, therefore, we simply add the coefficients 3 and 4 giving an answer of 7. The powers stay the same. You do not add or subtract the powers.

Example	Explanation
2. $6 \times 10^4 + 3 \times 10^5$ $= 6 \times 10^4 + 3 \times 10^1 \times 10^4$ $= 6 \times 10^4 + 30 \times 10^4$ $= 36 \times 10^4$	In example 2, you <u>cannot</u> add the coefficients because the powers are *not* the same. Therefore, you must make the powers the same and rewrite this equation in its modified form. In order to achieve this, we need to reduce the power 5 by one in the second term or increase the power 4 by one. In this example, you split the 10^5 as $10^1 \times 10^4$ by using the additive rule for addition of powers when the base is the same. Then multiply 3 and 10^1 which will give you 30. At this point, you can add the coefficients 6 and 30 giving rise to 36, keeping powers the same.
3. $5 \times 10^6 - 20 \times 10^5$ $= 5 \times 10^1 \times 10^5 - 20 \times 10^5$ $= 50 \times 10^5 - 20 \times 10^5$ $= 30 \times 10^5$	Here again in example 3, the subtraction is tricky because the powers are not the same for base 10. In this case we split the power 10^6 as $10^1 \times 10^5$ and multiply the coefficient of this term, 5, by 10^1 ($5 \times 10 = 50$). Write the equation in its modified form. Now the powers are the same and, therefore, we can subtract the coefficients. If you do that, we will get an answer of 30, keeping the powers of 10 the same.
4. $32 \times 10^{-5} - 20 \times 10^{-6}$ $= 32 \times 10^{-5} - 20 \times 10^{-1} \times 10^{-5}$ $= 32 \times 10^{-5} - 2 \times 10^{-5}$ $= 30 \times 10^{-5}$	In this example, you split 10^{-6} as $10^{-1} \times 10^{-5}$ first. Then divide the coefficient of the second term by 10 (which is the same thing as multiplying with 10^{-1}), e.g., $20/10 = 2$. Write the equation in its modified form. Now subtract the coefficients ($32 - 2 = 30$).

Exercise Set 9.3

Simplify the indicated operations.

These operations are quite challenging. Be prepared to take time with the students and have a boatload of patience. Have several students come to the chalkboard to show their solutions to each of the problems. Discuss the solutions and the steps involved with these problems, and involve other students in the class. Don't rush.

1. $4 \times 10^4 + 5 \times 10^4 =$
2. $4 \times 10^5 + 5 \times 10^4 =$
3. $4 \times 10^{29} + 2 \times 10^{29} =$
4. $4 \times 10^5 + 5 \times 10^6 =$
5. $3 \times 10^9 - 15 \times 10^8 =$
6. $3 \times 10^{11} - 50 \times 10^9 =$
7. $512 \times 10^{-5} - 120 \times 10^{-6} =$
8. $3 \times 10^9 - 30 \times 10^8 =$
9. $512 \times 10^{-5} - 1120 \times 10^{-6} =$
10. $256 \times 10^{-15} - 560 \times 10^{-16} =$

SCIENTIFIC NOTATION

Large numbers (e.g., 3,573,000) and small numbers (0.000267) are often written in scientific notation. To write a given number, such as these two, in scientific notation, it must be a product of *two* factors. The *first* factor is a number between 1 and 10, including 1; the *second* factor is a power of 10.

For example, to write 3,573,000 in scientific notation, we must first position the decimal point to get a number between 1 and 10. This means that we must place a decimal after the first number 3 and before the number 5.

In the original whole number, the decimal point is assumed to be at the end of the number, after the last 0 although it is not stated. After writing the same number in scientific notation, the two numbers (the original number and the other one written in scientific notation) must equal each other. In order to place a decimal point between 3 and 5 as required by the scientific notation pronciple, we had to shift the decimal point to the left by six places. The number of places (in this case, 6) by which a decimal point needs to be shifted to the left to obtain the first factor between 1 and 10 then becomes the power of 10.

We now have the two factors. The first is the number 3.573000 and the second is 10^6. We then write the given number as a product of the two factors, which gives us 3.573000×10^6 or simply 3.573×10^6, the scientific notation for 3,573,000.

- *Rule 1:* When the decimal place is shifted to the *left* to obtain the first factor for the scientific notation, the number of places by which the decimal place needs to be shifted gives a *positive* power for 10.

Another example is to a write smaller number, like 0.000267, in scientific notation. Again the task here is the same as before, that is, to write this number as two factors. The first factor has to be between 1 and 10. Therefore, the decimal must be placed between 2 and 6. This will require moving the decimal place in the original number four places to the *right*. The shifting of the decimal place four places gives the second factor, 10^{-4}.

The two factors here are 2.67 and 10^{-4}. Therefore, we can write the number 0.000267 as a product of these two factors, i.e., 2.67×10^{-4}, the scientific notation for 0.000267.

- *Rule 2:* When the decimal place is shifted to the *right* to obtain the first factor for the scientific notation, the number of places by which the decimal place needs to be shifted gives a *negative* power for 10.

Exercise Set 9.4

Write the following into scientific notation.

Involve every student in providing answers to these questions. The answers require no writing, paper, or pencil. Instead, write these questions on the chalkboard, and then have particular students called upon to answer these questions. With the more challenging questions, have more than one student answer the question. If the answers vary with these questions, make certain to explain why one answer is right and another is not. Then make sure every student has the correct answer.

1. 24,000
2. 500,000
3. 4,480,000
4. 567,000
5. 23,000,000
6. 62,000
7. 80
8. 700
9. 9,000
10. 9,345,000
11. 600,000,000
12. 7,200,000,000,000
13. 0.0012
14. 0.000135
15. 0.000354
16. 0.0267
17. 0.568
18. 0.000523
19. 0.000000879
20. 0.0000000489
21. 0.00829
22. 0.0389
23. 0.00000065
24. 0.000000000089

Exercise Set 9.5

Write each number in the following statements in scientific notation.

It is a good idea for students to see numbers written out, especially when they are long. Therefore, write only the numbers that are long on the chalkboard, and students need to provide an answer mentally. For short numbers, just say them.

1. Lightning from the clouds strikes the Earth approximately 8,670,000 times every day.
2. The average body mass in grams is approximately 74,000.
3. A flight from Washington, D.C., to New York usually takes around 3600 seconds (an hour).
4. In 2004, a computer typically took approximately 0.00000000045 seconds to complete one instruction.
5. Light will take approximately 0.032 seconds to make a round-trip between Washington, D.C., and California.

MENTAL CHAPTER REVIEW

Lead the class into this chapter review using the pedagogical skills of the students' differential knowledge bases and equal participation at every step. As a first step, ask the students to write the following in scientific notation on a piece of paper. When the students have completed the task, ask them by name for the answer. If there is an error in the response, have another student help out by explaining the problem-solving process.

Write in scientific notation:

1. 10
2. 16
3. 467
4. 9823
5. 56,000,000
6. 567,234,000
7. 7,400,000,000
8. 95
9. 1
10. 23.34
11. 99
12. 0.1
13. 0.23
14. 0.589
15. 0.004589
16. 0.000056
17. 0.0000491
18. 0.00000089
19. 0.0000000045
20. 0.000783
21. 0.0000000001

Perform the indicated operations, and write the final answer in scientific notation.

1. $10^1 \times 10 =$
2. $10^{12} \times 10^{12} =$
3. $10^{13} \times 10^{13} =$
4. $10^{16} \times 10^{16} =$
5. $10^{20} \times 10^{15} =$
6. $10^{-10} \times 10^{-10} =$

7. $10^{-32} \times 10^{-32} =$

8. $10^{-15} \times 10^{-10} =$

9. $10^{14} \times 10^{-4} =$

10. $10^{12} / 10^{-10} =$

11. $10^{14} / 10^{-10} =$

12. $10^{5} / 10^{-10} =$

13. $10^{12} / 10^{13} =$

14. $10^{-1} / 10^{-1} =$

15. $10^{-16} / 10^{-16} =$

16. $40 \times 10^4 + 50 \times 10^4 =$

17. $5 \times 10^5 + 6 \times 10^5 =$

18. $6 \times 10^5 + 7 \times 10^4 =$

19. $6 \times 10^{22} + 4 \times 10^{22} =$

20. $50 \times 10^5 + 5 \times 10^6 =$

21. $4 \times 10^5 + 60 \times 10^4 =$

22. $3 \times 10^9 - 5 \times 10^8 =$

23. $2 \times 10^{11} - 50 \times 10^9 =$

24. $3 \times 10^5 + 70 \times 10^4 =$

25. $256 \times 10^{-5} - 560 \times 10^{-6} =$

26. $3 \times 10^9 - 30 \times 10^8 =$

27. $4 \times 10^5 - 40 \times 10^4 =$

28. $30 \times 10^9 - 100 \times 10^8 =$

29. $4 \times 10^{11} - 200 \times 10^9 =$

30. $3 \times 10^7 + 500 \times 10^4 =$

SAMPLE EXIT QUESTIONS

Simplify:
1. 212 cm = _____ m
2. $\dfrac{10^2 * 10^{-5} * 10^{-7}}{10^3 * 10^{-13}} =$
3. Write into scientific notation after simplifying: $2 \times 10^6 + 4 \times 10^5 + 6 \times 10^5$

SAMPLE PRIMING HOMEWORK

Simplify:

1. $(x^3)(x^9) =$

2. $(x^{11})(x^{-9}) =$

3. $(y^{-11})(y^9) =$

4. $(y^{-11})(y^{-9}) =$

5. $\dfrac{x^9}{x^9} =$

6. $\dfrac{x^9}{x^{-9}} =$

7. $\dfrac{x^{-9}}{x^{-9}} =$

8. $\dfrac{x^{-9}}{x^9} =$

POINTS TO REMEMBER AND REVIEW

- In scientific notation, the power of 10 will be *positive* if the original number is more than 10 and *negative* if the number is less than 1.
- When adding or subtracting terms involving coefficients and powers of 10, make the power of 10 the same for all terms before adding or subtracting their coefficients.
- Remember the rules for adding negative numbers: if all of the numbers are negative, add all of them, and place a *negative sign* in front of the answer.
- Overlearn the rules of manipulating powers of 10 and of exponents in general.
- In converting from one unit to another, first visualize what the answer will be. If the answer you expect is more than the original number, multiply the original number with the conversion factor. If the answer is going to be less than the original number, divide the original number by the conversion factor. For example, 2.3 m = 230 cm. In this case, the original number 2.3 is multiplied by the conversion factor 100 (1 m = 100 cm) to get an answer of 230 because the answer is expected to be more than 2.3.

10
Powers of Variables and Constants

In scientific work, we use not only the powers of 10 as explained in the previous chapter but also the powers of variables ($x, y, z, a, b, c, p, q, r, s$, etc.) and of constants. A *variable* is a term represented by a symbol that does not have a fixed value, whereas a *constant* is something that is fixed, like 11 or π.

To illustrate this point, let us take a square with a side of 6 inches long. The area (length × width) of this *square* is $6 \times 6 = 6^2 = 36$ in^2. We read the term 6^2 as "6 squared" or "6 to the second power." In another illustration, the *volume* (length × width × height) of a cube that has a side of 4 inches will be $4 \times 4 \times 4 = 4^3 = 64$ in^3. The term 4^3 is read as "4 cubed" or "4 to the third power." In these examples, we are dealing with the powers of constants.

Writing the same problems in terms of variables, we can use x as the side of a square. Then, the area of a square will be x^2. Similarly, if the side of a cube is p, then the volume of the cube will be p^3. In these cases, we are using variables rather than constants for the side of a square or the side of a cube, which is quite common in any scientific work.

Tell the students to remember that instead of writing the same factor—which can be either a variable or a constant—several times, they can write the factor once and then use an exponent to indicate the number of repetitions. For instance, $3 \times 3 \times 3 \times 3 = 3^4$, three multiplied by itself four times.

Many times in our work, we replace the multiplication sign (×) with an elevated period (·) or an asterisk (*). Using an asterisk for multiplication is very common in computer-related work. To show how you can extend the students' understanding of the examples above, we write expressions like the following ones for our daily work in math and science.

- $3 \cdot 3 \cdot 3 \cdot 3 = 3^4$ or 3 to the fourth power (3 multiplied with itself four times)
- $4 \cdot 4 \cdot 4 \cdot 6 \cdot 6 \cdot 6 \cdot 6 \cdot 6 = 4^3 \cdot 6^5$ (4 multiplied by itself 3 times, and 6 by itself five times)

In an algebra class, we will use variables instead of constants. For example:

- $n \cdot n \cdot n \cdot n = n^4$ (n multiplied by itself four times)
- $p \cdot p \cdot p \cdot p \cdot q \cdot q = p^4 q^2$ (p multiplied by itself four times and q by itself twice)
- $3 \cdot a \cdot a \cdot 4 \cdot b \cdot b \cdot b = 12 a^2 b^3$ (multiplying 3 with 4 gives 12; a is multiplied by itself two times and b by itself three times)

However, before we proceed with the task of powers of variables, we will review the concepts learned in the previous chapters. These concepts help with the learning of the powers of variables.

REVIEW OF PREVIOUS KNOWLEDGE

Ask your students to provide the answer to all the problems after doing them mentally without pen and paper and before moving on to the next section in this chapter.

Review of Basic Math Operations
1. Write a word name for 613,011.
2. Write a word name for 7,607,020.
3. Write a number for six million, two hundred ninety thousand, and three hundred forty.

Simplify:

4. 460 + 540 =

5. 996 + 1004 =

6. 3056 − 1006 =

7. 4099 − 2099 =

8. 14 * 14 =

9. 11 * 11 =

10. 64 / 8 =

11. 100 / 10 =

Powers of Variables and Constants

12. $7^2 =$
13. $11^2 =$
14. $\sqrt{36} =$
15. $\sqrt{100} =$
16. $(49)^{1/2} =$
17. $(64)^{1/2} =$
18. $1234 / 0 =$
19. $\frac{1}{3} \div 3 =$
20. $\frac{8}{2} \times \frac{6}{3} =$
21. $\frac{129}{2} + \frac{129}{2} =$
22. $\frac{11}{16} \div \frac{11}{16} =$
23. $14 * 0 =$
24. $\frac{0}{67} =$
25. $\frac{11}{0} =$
26. $\frac{7}{0} =$
27. $\frac{0}{67} =$
28. $\frac{100}{0} =$
29. $\frac{0}{56} =$
30. $\frac{54}{0} =$
31. $12^0 =$
32. $(y^0) =$
33. $(2y)^0 \times (1000)^0 =$
34. $(2v)^0 =$

Write the following in scientific notation.

1. $1 =$
2. $10 =$
3. $680 =$
4. $4096 =$
5. $256{,}000{,}000 =$
6. $2{,}800{,}000{,}000 =$
7. $0.0982 =$
8. $0.0005269 =$
9. $456.0089 =$

Perform the indicated operations and simplify.

1. $10^0 \times 1 \times 100^0 =$
2. $10 \times 10^2 =$
3. $10^6 \times 10^7 \times 10^{17} =$
4. $10^{19} \times 10^{19} =$
5. $10^{-32} \times 10^{-32} =$
6. $10^{-15} \times 10^{-15} =$
7. $10^{17} / 10^{-10} =$
8. $10^{13} / 10^{20} =$
9. $10^{-6} / 10^{-6} =$
10. $10^{-17} / 10^{-17} =$
11. $10 \times 10^6 + 15 \times 10^6 =$
12. $91 \times 10^8 + 9 \times 10^8 =$

Exercise Set 10.1

Write the following using exponents.

1. $8 \cdot 8 \cdot 8 =$
2. $6 \cdot 6 \cdot 6 \cdot 6 \cdot 6 =$
3. $2 \cdot 2 \cdot 2 \cdot 5 \cdot 5 =$
4. $7 \cdot 7 \cdot 8 \cdot 8 \cdot 8 \cdot 6 \cdot 6 \cdot 6 \cdot 6 =$
5. $x \cdot x \cdot x \cdot x =$
6. $x \cdot x \cdot x \cdot y \cdot y =$
7. $p \cdot p \cdot p \cdot q \cdot q \cdot q \cdot q =$
8. $m \cdot m \cdot 3 \cdot n \cdot n \cdot n \cdot 5 =$
9. $6 \cdot a \cdot a \cdot 4 \cdot a =$
10. $2 \cdot b \cdot b \cdot c \cdot c \cdot c \cdot 6 =$
11. $r \cdot r \cdot 3 \cdot s \cdot r \cdot 5 \cdot s =$
12. $8 \cdot x \cdot y \cdot x \cdot y \cdot y \cdot 9 \cdot x =$
13. $c \cdot d \cdot d \cdot c \cdot 9 \cdot 9 \cdot d \cdot c =$
14. $2 \cdot b \cdot c \cdot d \cdot 4 \cdot d \cdot 7 \cdot b =$
15. $t \cdot 3 \cdot 3 \cdot d \cdot v \cdot v \cdot t \cdot 3 \cdot d =$

Powers of Variables and Constants 141

MATHEMATICAL OPERATIONS INVOLVING EXPONENTS

Multiplication Involving Like Bases

In multiplication when the base is the same, you simply add the powers. For example:

Example 10.1. Multiplying Terms with the Same Base and Different Powers

	Example	Explanation
1.	$x^2 \cdot x^3 = x^5$	Here the base, which is x, is the same in both cases. Therefore, the final answer of multiplication is x^5 which is obtained by simply adding the powers (2 and 3) of similar bases.
2.	$a^5 \cdot a^{-3} = a^2$	Here again the bases are the same. Therefore, we add the powers (5 and –3) giving rise to (5 + –3) = 2 for the answer.

Exercise Set 10.2

Simplify the following.

  Students need to simplify the following expressions mentally with whatever support is needed. Write these problems, one at a time, with big and spaced-out letters on the chalkboard. In seeking answers from students, involve all of them by calling on them by their first names. Give the students time to respond, and allow them to write the answers down in their notes if they want to do so.

1. $a \cdot a^2 \cdot a^4 =$

2. $b \cdot b^2 \cdot b^2 \cdot b^2 \cdot b^2 =$

3. $x^{-8} \cdot x^2 \cdot x^2 \cdot x^2 \cdot x^2 =$

4. $y^6 \cdot y^{-3} \cdot y^{-3} =$

5. $m^5 \cdot m^5 \cdot m^{-10} =$

6. $x^2 \cdot x^4 \cdot x^{-2} \cdot x^{-2} =$

7. $x^9 \cdot x^8 =$

8. $p^{10} \cdot p^{-10} =$

9. $m^2 \cdot m^3 \cdot n^3 \cdot n^4 =$

10. $a^4 \cdot b^3 \cdot a^4 =$

11. $b^3 \cdot b^5 \cdot c^2 \cdot c^3 \cdot c^5 =$

12. $r^5 \cdot r^5 \cdot 3 \cdot s^4 \cdot r^5 \cdot 5 \cdot s^4 =$

13. $8 \cdot x^2 \cdot y^3 \cdot x^2 \cdot y^3 \cdot y^2 \cdot 9 \cdot x^5 =$

14. $c^2 \cdot d^3 \cdot d^3 \cdot c^2 \cdot 8 \cdot 8 \cdot d^3 \cdot c^2 =$

15. $2 \cdot b^3 \cdot c^2 \cdot d^4 \cdot 4 \cdot d^4 \cdot 7 \cdot b^3 =$

16. $t^2 \cdot 3 \cdot 3 \cdot d^5 \cdot v^6 \cdot v^6 \cdot t^2 \cdot 3 \cdot d^5 =$

Division Involving Like Bases

When the bases of terms are the same and you are dividing, *subtract* the powers. For example:

Example 10.2. Dividing Terms with the Same Base and Different Powers

	Example	Explanation
1.	x^5 / x^3 $= x^2$	Here the two bases (x and x) are the same. Therefore, the final answer of this division is x^2. You obtain this answer by subtracting the powers 5 and 3, e.g., $(5 - 3 = 2)$, and keep the base as is.
2.	a^6 / a^{-2} $= a^{6+2}$ $= a^8$	Here again the bases (a and a) are the same. Therefore, we subtract the powers 6 and -2, e.g., $[6 - (-2)]$ giving rise to $(6 + 2) = 8$ for the answer as the power of a.

Another approach commonly used in division involving powers of variables and constants with the same bases is to bring the quantities with exponents from the denominator to the numerator. In doing so, the exponent of that quantity changes signs. Afterward, you add the powers of the quantities with the same bases. Let us illustrate it with two other examples:

Example 10.3. Dividing Terms with the Same Base and Different Powers, continued ...

	Example	Explanation
1.	$p^{12} / p^9 = p^{12} \cdot p^{-9}$ $= p^{12-9} = p^3$	First, we bring p^9 from denominator to numerator (reciprocal operation). In doing so, you change the sign of the power from 9 to –9. Then we added the two powers $[12 + (-9)]$ of p. The answer for the power of p is, therefore, 3. Remember, $\frac{1}{p^9} = p^{-9}$, therefore, $\frac{p^{12}}{p^9} = p^{12} \cdot p^{-9} = p^{12+(-9)} = p^3$.
2.	$n^8 / n^{-7} = n^8 \cdot n^7$ $= n^{8+7} = n^{15}$	Again, we bring from the denominator to numerator (reciprocal operation) changing the sign of the power from –7 to +7. Then, we treat this problem as a product with the same bases. Therefore, we added the two powers (8 and 7), giving rise to an answer of 15 for the power of n.

Exercise Set 10.3

Simplify the following expressions mentally.

You should write these problems on the chalkboard before asking a student to answer the question. Use the principle of students' differential knowledge bases to seek answers. This practice ensures full participation of the class.

1. $\dfrac{a^6}{a^4} =$

2. $\dfrac{b^{19}}{b^9} =$

3. $\dfrac{x^{13}}{x^{13}} =$

4. $\dfrac{y^6}{y^{-3}} =$

5. $\dfrac{x^2}{x^{12}} =$

6. $\dfrac{x^9}{x^{13}} =$

7. $\dfrac{p^{10}}{p^{-10}} =$

8. $\dfrac{m^{-16}}{m^{-16}} =$

9. $\dfrac{a^4 \cdot b^3}{a^4} =$

10. $\dfrac{c^5 \cdot b^3}{b^2 \cdot c \cdot c^4} =$

11. $\dfrac{r^5 \cdot r^5 \cdot 8 \cdot s^5}{r^5 \cdot 4 \cdot s^4} =$

12. $\dfrac{16 \cdot x^2 \cdot y^3 \cdot x^2 \cdot y^3}{y^2 \cdot 8 \cdot x^5} =$

13. $(c^2 \cdot d^3) / (d^3 \cdot c^2 \cdot 2 \cdot 4 \cdot d^3 \cdot c^2) =$

14. $(28 \cdot b^3 \cdot c^2) / (d^4 \cdot 4 \cdot d^4 \cdot 7 \cdot b^3) =$

Addition and Subtraction Involving Like Bases

This is an extremely important concept in math and the sciences, but as teachers ourselves, we have witnessed that many students make mistakes when dealing with this type of addition and subtraction. When adding or subtracting in expressions that involve exponents, you add or subtract the *coefficients* of these exponents *only when the variables are the same and they each have the same power.* If the powers of the variables are not the same or the variables themselves are not the same, leave those exponents alone. For example:

Example 10.4. Collecting Like Terms

Example	Explanation
1. $2p^3 + 3p^3 = 5p^3$	In an expression like this, the coefficients are 2 and 3 of the quantities or variables with the same power. Therefore, only add the coefficients, the final answer for the coefficients alone will be 5. <u>Do not</u> add the powers of the variables, in this case p. The answer to the problem is $5p^3$.
2. $9q^5 - 6q^5 = 3q^5$	Here again the powers of q are the same. Therefore, we subtract the coefficients that are 9 and 6, giving the answer of 3 for the subtraction. The final answer will be $3q^5$.
3. $6x^4 + 5x^4 - 3x^4 + 12x^5$ $= 8x^4 + 12x^5$	An addition problem like this is at best tricky. Here we only combine terms when the powers of x are the same, i.e., add the coefficients 6, 5, and -3 to get an answer of 8 as a single coefficient of x^4. Leave the other term with x^5 alone. It is altogether different than x^4.
4. $8z^7 + 6z^7 - 3z^6 + 14z^6$ $= 14z^7 + 11z^6$	Here again, add or subtract the coefficients of terms having the same powers of the variables. In this example, we have two terms with z^7 and another two terms with z^6. Now combining similar terms, we will have the coefficient of 14 <u>after</u> adding 8 and 6 as the final coefficient for z^7. Similarly, we will have an answer of 11 after adding 14 + (-3) as the coefficient of z^6.

Exercise Set 10.4

Simplify the following expressions.

    Write these expressions on the chalkboard clearly with big letters and plenty of spacing. Let students copy them on their papers before they provide the final answer. Exercises like this are time consuming. Let students take time to answer. Ask some students to come to the chalkboard to ensure their understanding of these concepts fully. Practice is the only way to get the hang of these problems. Check everyone's notes to make certain each student has performed the operations.

1. $3p^3 + 6p^3 =$
2. $7p^6 + 6p^6 =$
3. $p \cdot q + q \cdot p + 3p \cdot q =$
4. $m \cdot p + 2p \cdot m + 3p \cdot m - 5p \cdot m =$
5. $11x \cdot y + 11y \cdot x + 13q^2 + 14q^2 \cdot p =$
6. $14p \cdot r + 16r \cdot p + 30p \cdot r =$
7. $7p^6 + 6p^6 + 3p^6 + 4p^6 =$
8. $17p^5 + 16p^5 + 13p^5 + 14p^5 =$
9. $11q^5 - 4q^5 =$
10. $11q^5 - 4q^5 - 5q^5 - 2q^5 =$
11. $9x^4 + 5x^4 - 13x^4 + 7x^7 =$
12. $9z^7 + 9z^7 - 3z^6 + 14z^6 - z^6 =$
13. $p^2 + 2p^2 + 3p^2 + 2q^3 + 3q^3 + 4q^3 + 6r^4$
14. $p^2 + 2p^2 - 3p^2 + 2q^3 + 3q^3 - 5q^3 + 6r^4$
15. $m^3 + 2m^3 - 4m^3 + 2q^3 + 3q^3 - 7q^3 + n^4 + 9n^4$
16. $m^3 - 4m^3 + 2q^3 + 3q^3 + 4n^4 + 2m^3 - 7q^3 - 6n^4$
17. $32m \cdot n^2 + 32n^2 \cdot m + 64m \cdot n^2 + 12q$
18. $p \cdot q^3 + q^3 \cdot p + 8p \cdot q \cdot q^2 + 10p \cdot q \cdot q \cdot q$
19. $3t^2 \cdot u^4 + 3u^4 \cdot t^2 + 4t \cdot t \cdot u^2 \cdot u^2$
20. $9m^2 \cdot n^2 \cdot p^3 + 9m \cdot m \cdot n \cdot n \cdot p^2 \cdot p + 9m^2 \cdot n \cdot n \cdot p \cdot p^2 + 11m^2 \cdot n^2 \cdot p^3 \cdot q$

CONSTANTS AND VARIABLES WITH A POWER RAISED TO A POWER

In exponential notation, when a quantity or a variable is raised to a power and then *that* power is raised to another power again to get a final answer, you simply multiply the powers. The rules of multiplication, division, addition, and subtraction you have previously covered continue to apply here as well. For example:

Example 10.5. Terms with Power Raised to Another Power

Example	Explanation
1. $(4^2)^3 = 4^2 \cdot 4^2 \cdot 4^2$ $= 4^{2+2+2}$ $= 4^6$	First, the expression is written in long hand, i.e., 4^2 is multiplied by itself three times. The powers of 4 are then added together to give an answer of 6 for the power. A short cut to this long approach is to multiply the power 2 with the power 3 to obtain an answer of 6.

Example	Explanation
2. $(5^3)^4$ $= 5^{3 \cdot 4}$ $= 5^{12}$	The base is a constant, but you get the final answer again by multiplying the powers 3 and 4, which is 12. The final answer, therefore, is 5^{12}.
3. $(3b^3)^2$ $= 3^2 b^6$ $= 9b^6$	Here the expression $3b^3$ is raised to another power of 2. So, the problem is in two parts. First deal with the 3 within the parentheses and then with b^3, also in the parentheses. The powers of 3 are 1 and 2, so we get 2 (product of 2 and 1) for the power of 3. Similarly, the powers of b are 3 and 2. If we multiply 3 and 2, we get an answer of 6 for the power of b. The final answer, therefore, is $3^2 b^6 = 9b^6$.
4. $(y^3)^3 / (y^4)^2$ $= y^9 / y^8$ $= y^{9-8}$ $= y$	For the numerator, we get y^9 after multiplying powers 3 and 3. For denominator, we get y^8 once we multiply powers 4 and 2. Then we apply the rule of division for powers when bases are the same, i.e., subtract the powers. After subtracting 8 from 9, we will get 1 for the power of y.

Exercise Set 10.5

Simplify the following expressions.

Write the following expressions on the chalkboard before soliciting an answer from the students. Make certain these expressions are clearly written and appropriately spaced for students to see clearly what is expected of them in arriving at a correct answer. This set of tasks is the meat of any algebra course.

1. $(3^2)^3 =$
2. $(6^5)^3 =$
3. $(9^3)^9 =$
4. $(5^{1/2})^2 =$
5. $(7^{1/2})^4 =$
6. $(3^{1/2})^6 =$

7. $(b^4)^2 =$
8. $(m^3)^4 =$
9. $(3y^2)^2 =$
10. $(9z^6)^2 =$
11. $(3z^5)^3 =$
12. $(2x^2 y^3)^2 =$

13. $(4m^3 n^5)^2 =$
14. $(p^4 q^6)^3 =$
15. $(z^4)^3 / (z^3)^3 =$
16. $(y^5)^3 / (y^3)^5 =$
17. $(m^4 \cdot m^3)^3 =$
18. $(p^3 q^6)^3 / (p^2 q^3)^6 =$

FRACTIONAL POWERS OF CONSTANTS AND VARIABLES

The rules of multiplication, division, addition, and subtraction involving fractional powers of variables are the same as for the whole number powers of variables. Provide students ample time to complete these exercises. We will illustrate these rules with the help of examples:

Example 10.6. Multiplying, Dividing, and Collecting Like Terms

Example	Explanation
1. $5^{1/2} \cdot 5^{1/2}$ $= 5^{1/2 + 1/2}$ $= 5^1$ $= 5$	The base, a constant, is the same. Therefore, in multiplying terms with the same base, we add the powers. The addition, 1/2 plus 1/2 equals 1. Base being the same, 5, the final answer is 5^1, which is the same as 5.
2. $m^{1/2} \cdot m^{1/2}$ $= m^1$ $= m$	The powers of m, the base, are 1/2 and 1/2. The sum of these two numbers is simply one. Therefore, the base, m, has a power of 1 for the final answer.
3. $n^{1/3} \cdot n^{1/3} \cdot n^{1/3}$ $= n^{1/3 + 1/3 + 1/3}$ $= n^{3/3} = n^1 = n$	Here again, if we add 1/3 by itself three times, the sum will be 1. The power of n, therefore, is again 1.
4. $y^{1/2} / y^{1/2}$ $= y^{1/2 - 1/2}$ $= y^0$ $= 1$	When dividing powers that involve the same base, you subtract the powers. Therefore, if you subtract 1/2 from 1/2, you get zero. The answer is 1 because any variable or constant that is raised to power of 0 gives an answer of 1.
5. $p^{1/2} + 2\,p^{1/2} + 3\,p^{1/2}$ $= (1 + 2 + 3)\,p^{1/2}$ $= 6\,p^{1/2}$	Here the powers and the base are the same. Therefore, you add the coefficients to get an answer. The sum of the coefficients is: $1 + 2 + 3 = 6$. The complete answer, therefore, is $6\,p^{1/2}$.
6. $6\,p^{1/3} - 2\,p^{1/3} + 4\,p^{1/3}$ $= (6 - 2 + 4) = 8\,p^{1/3}$	The answer of 8 for the coefficient is obtained by adding the coefficients $6 - 2 + 4 = 8$ of the individual terms.

Example	Explanation
7. $6q^{1/3} - 2q^{1/3} + 8q^{1/4} - 3q^{1/4}$ $= (6-2)q^{1/3} + (8-3)q^{1/4}$ $= 4q^{1/3} + 5q^{1/4}$	This example has the same base with different powers. Therefore, collect like terms first before adding the coefficients of these terms. The first two terms are similar because they have the same base and the same powers. Therefore, we add the coefficients 6 and -2 to get an answer of 4. Similarly, the last two terms are like terms, having the same base and powers. To simplify the last two terms, add their coefficients 8 and -3 and obtained an answer of 5.

Exercise Set 10.6

Simplify the following expressions.

Provide a handout of these problems instead of writing them on the chalkboard. (Having the students copy the problems is both time consuming and filled with the potential for mistakes.) Instruct students to leave the answer as a power of a constant and/or a variable if the simplification of the final expression is difficult.

1. $(3^{1/2})^4 =$
2. $(6^{1/5})^{10} =$
3. $(9^{2/5})^5 =$
4. $(b^{4/2})^{1/2} =$
5. $(m^{3/4})^4 =$
6. $(3y^{1/2})^2 =$
7. $(16z^{6/3})^{1/2} =$
8. $(z^{1/5})^{15} =$
9. $(4x^2y^4)^{1/2} =$
10. $x^{1/3} + 3x^{1/3} + 4x^{1/3} =$
11. $8y^{1/3} - 10y^{1/3} + 3y^{1/3} =$
12. $8q^{1/3} - 5q^{1/3} + 3q^{1/4} =$
13. $m^{2/3} / m^{2/3} =$
14. $m^{4/3} / m^{1/3} =$
15. $p^{7/3} / p^{1/3} =$
16. $x^{1/3} + 7x^{1/3} + 6x^{1/3} =$
17. $8y^{1/3} - 12y^{1/3} + 4y^{1/3} =$
18. $10q^{1/3} - 6q^{1/3} + 4q^{1/4} =$
19. $x^{1/2} + 5x^{1/2} + 4x^{1/2} =$
20. $5y^{1/2} - 10y^{1/2} + 5y^{1/2} =$
21. $5q^{2/3} - 3q^{2/3} + 5q^{3/4} =$
22. $m^{5/2} / m^{1/2} =$
23. $x^{1/3} + 5x^{1/3} + 14x^{1/3} =$
24. $18q^{1/2} - 12q^{1/2} + 2q^{1/3} =$

Powers of Variables and Constants 149

MENTAL CHAPTER REVIEW

Call every student by name using the principle of differential knowledge bases to test students' understanding. Write every problem on the chalkboard before asking a question. Ask students to come to the chalkboard to drive the point home about these problems.

Simplify the following.

1. $x^2 \cdot x^2 \cdot x^{-2} \cdot x^{-2} =$
2. $x^9 \cdot x^9 =$
3. $p^{15} \cdot p^{-15} =$
4. $m^3 \cdot m^3 \cdot n^4 \cdot n^4 =$
5. $y^6 \cdot y^{-3} \cdot y^{-3} =$
6. $x^9 \cdot x^7 =$
7. $x^3 / x^{12} =$
8. $x^8 / x^{13} =$
9. $p^{10} / p^{-10} =$
10. $m^{-16} / m^{-16} =$
11. $(b^4 \cdot c^5) / (b^2 \cdot c \cdot c^4) =$
12. $(a^3 \cdot b^3) / a^3 =$
13. $7p^6 + 6p^6 + 3p^6 =$
14. $17p^5 + 16p^5 + 13p^5 =$
15. $9x^4 + 7x^4 - 13x^4 + 7x^7 =$

16. $11q^5 - 4q^5 =$
17. $9z^7 + 9z^7 - 3z^6 + 14z^6 =$
18. $11q^5 - 4q^5 - 5q^5 - 2q^5 =$
19. $(5^{1/2})^2 =$
20. $(7^{1/2})^4 =$
21. $(3^{1/2})^6 =$
22. $(b^4)^3 =$
23. $(m^3)^4 =$
24. $(3y^2)^3 =$
25. $(b^{4/2})^{1/2} =$
26. $(m^{3/4})^4 =$
27. $(3y^{1/2})^4 =$
28. $(4z^{6/3})^{1/2} =$
29. $(z^{1/5})^{10} =$
30. $(4x^2y^4)^{1/2} =$

SAMPLE EXIT QUESTIONS

Simplify.

1. $(x^{15})^{1/5} =$

2. $\dfrac{x^6}{x^{-9}} =$

3. $3(x^9)^{1/3} + 4(x)(x^2) - 9\dfrac{x}{x^{-2}} =$

4. $\sqrt{64x^8} + 2(x)(x^3) - (9x^8)^{1/2} - \dfrac{24x^6}{2x^2} =$

SAMPLE PRIMING HOMEWORK

1. Write an expression for "The product of p and q minus 5 equals 11."
2. A number increased by 8 equals 11, what is the number?
3. Solve for x: $3x + 2x + 6x = 33$
4. Solve for y: $3y - 5 = -2y + 10$
5. Solve for z: $2(-3z + 5) = 3(z - 4 - 2) + 1$

POINTS TO REMEMBER AND REVIEW

- Overlearn the general formulas in dealing with powers of exponents:

  $x^m \cdot x^n = x^{m+n}$

  $(x^m)^n = x^{mn}$

  $\dfrac{x^m}{x^n} = x^{m-n}$

  $x^{-1} = \dfrac{1}{x}$

  $x = \dfrac{1}{x^{-1}}$

- When bringing terms from the denominator to the numerator or vice versa, change the sign of the power of the term in question.
- It is understood that x^1 is the same as x.

11
The World of Expressions and Elementary Equations

By definition, an *expression* does not have an equal sign, whereas an *equation* does. An equation typically asks you to solve for an unknown value, whereas an expression only asks you to evaluate and then simplify it. For example, $3n + 7$ is an expression (no equal sign). In contrast, $3n + 7 = 16$ (has an equal sign) is an equation, with n being the unknown.

Students must master the necessary concepts in expressions and acquire a good foundation in the writing and simplification of expressions *before* proceeding with solutions of equations. Therefore, this chapter begins with developing and evaluating expressions.

EXPRESSIONS

Writing Expressions

Some examples of expressions involving multiplication and division are provided first.

Example 11.1. Practice in Writing Word Expressions into Algebraic Expression

Word expression	Algebraic expression
6 times a number p	$6p$
5 multiplied with a variable m	$5m$

Word expression	Algebraic expression
A variable y divided by 6	$\dfrac{y}{6}$
A variable z divided by 10 minus 9	$\dfrac{z}{10} - 9$
14 times a variable r plus 11	$14r + 11$

Exercise Set 11.1

Write an algebraic expression for each word problem.

Your students should be able to answer these questions mentally. You just need to read the problem to them. However, a handout for these problems might facilitate this process.

1. A variable *p* divided by 12.
2. Sum of a variable *z* and 9.
3. 6 times a variable *c*, minus 11.
4. 7 times a variable *h*, minus 8.
5. Product of 12 and *p*.
6. Sum of *x*, *y*, and *z*.
7. Difference of 14 and 5, times *p*.
8. Sum of 4 times *y* and 8 times *z*.
9. Sum of *p*, *q*, and *r*, minus 22.
10. Difference of 9 times *f* and 5 times *z*.

Evaluating Algebraic Expressions

This technique involves substituting numbers for variables and then evaluating and/or simplifying the resulting expression. To illustrate, we provide two examples.

Example 11.2. Evaluating an Expression

	Expression	Evaluating Expressions for $a = 3$, $b = 5$ and $c = 4$
1.	$5a + 6 + b$ $= 5(3) + 6 + 5$ $= 15 + 6 + 5$ $= 26$	Here we have a constants 6, variables *a* and *b*, and 5 as the coefficient of *a*. Substitute the values for *a* and *b*. Rewrite after multiplication. Add all the numbers.

2. $2(ab) + c/2$ Substitute the values for a, b, and c. [Remember
   $=2(3 \times 5) + 4/2$ to complete all math operations within the parentheses () <u>before</u> multiplying by the number outside it.]
   $=30 + 2$
   $=32$ Rewrite after multiplication and division
   Add the final numbers.

Exercise Set 11.2

Make certain the students do the actual substitutions in these expressions before they write the final answer. Have students come to the board to show their work. This will help to reinforce their learning.

Evaluate each expression for $a = 4$, $b = 8$, and $c = 9$.

1. $8a$
2. $\dfrac{b}{4}$
3. $10 + c$
4. $c - a$
5. $(8a) - (4b)$

6. $(ab) - 2c$
7. $3b - 6a$
8. $\dfrac{c}{3} + 3$
9. $\dfrac{ab}{4} + c$
10. $ac - 5b$

Exercise Set 11.3

Require the students to use two steps here: (1) writing the expression out, and (2) making the actual substitution. This exercise needs to be done on paper. Do not have the students attempt this mentally. Emphasize that "increase" means a plus (+) sign and "decrease" means a minus (−) sign.

Have your students write a word expression, and evaluate it for $g = 3$, $k = 5$, and $p = 20$.

1. p divided by k
2. g increased by p
3. 10 less than p
4. Product of g and k
5. Difference of p and k

6. p divided by 5, decreased by g
7. g times k, minus p
8. Difference of k and g
9. k divided by 5, minus 1
10. p divided by 10, minus 2

Combining Like Terms

Now we are going to build on what we have been doing with expressions. This time, we will combine terms that have the same variables, i.e., all of the x's, all of the p's, and so on, with their coefficients as constants. Let us consider the following examples. Walk the students through the steps of the first example. Then have them work the second example independently.

Example 11.3. Collecting Like Terms

Algebraic Expression	Collecting Like Terms
$6p + 3 + 5p + 6$ $= 6p + 5p + 3 + 6$ $= 11p + 9$	Collect terms that have p's only ($6p$ and $5p$) and constants (3 and 6).
$9x + 4x + 9 + 3 + 2x + 1$ $= 9x + 4x + 2x + 9 + 3 + 1$ $= 15x + 13$	Collect all like terms and simplify, i.e., collect all terms containing x and all terms that are constants only and then simplify.

Exercise Set 11.4

Simplify by combining like terms.
    Prepare handouts for these problems. Encourage students to group like terms first on paper before simplifying. Have them write out these steps for simplifying.

1. $4n + 3n + n$
2. $6b + 2b + 4b$
3. $11c + 5c + 3 + 11$
4. $3x + 2 + 5x + 5 + 2x + 3$
5. $5y + 6 + 4y + 6 - 9y - 12$
6. $12c + 4c + 4 + 16 + 4c$
7. $d + 2d + 5d + 11 - 7d - 10$
8. $12z + 5z - 12 - 2 - 14z + 16$
9. $x + y + z + 2x + 2z + 3y$
10. $x + 3y + 4z + 5 + 11 + 2y + 5z$

FIRST CUMULATIVE REVIEW

Prepare a handout of these problems for the students, and either let them complete the problems mentally or ask them to write out the answers on a piece of paper.

Write an algebraic expression for each word expression.

The World of Expressions and Elementary Equations

1. 14 more than a number y
2. 8 increased by a variable c
3. A variable m increased by itself
4. A variable p less 11
5. m plus a number k
6. 9 decreased by z
7. Difference of p and q
8. Difference of r and 3 times s
9. 6 times m, minus 11
10. 7 more than 5 times g

Evaluate each algebraic expression for $x = 4$, $y = 7$, and $z = 3$.

11. $x + 2y + 2$
12. $x + 11$
13. $z - 3$
14. $2y - 14$
15. $2x + 2y + 2z$
16. $x + z - y$
17. $x + 2y + 3z$
18. $2y - 2x - 2z$

19. $3y - 2x - 2y$
20. $4y - 4x$
21. $\dfrac{y}{7}$
22. $\dfrac{2z}{6}$
23. $\dfrac{3x}{12}$
24. $\dfrac{z}{3} + 9$
25. $\dfrac{2y}{7} + 5$
26. $6 + \dfrac{3x}{4}$

Simplify by combining like terms.

27. $x + y + z + 2x + 3y$
28. $p + 2p + 3 + 3p$
29. $m + 3 + 2m + 4 + 3n$
30. $x - 2y + 3z - 4 + 3 - x - 2y - 4z$
31. $p + q + r + 2p + 5$
32. $x^1 + x^2 + x^3 + 2x^1 + 3x^2 + 5$
33. $r^{20} + 3r^{20} - y^{20} + 3y^{20} - 5$
34. $x^2 + y^2 + z^2 + 2x^2 + 2y^2 + 4z^2$
35. $x + x^2 + x^3 + 3 + y^2 + 3y^3 + 4y^2$
36. $2p + 3x + 6z + 9 + 8p + 7x + 4z$

SIMPLE EQUATIONS

Now, having worked with expressions, we are going to move on to equations. Here are three aspects that students must familiarize themselves with when dealing with equations:

1. The definition of an equation
2. The disciplines that use equations
3. The techniques for solving equations

Let us begin with the first aspect: the definition. What is an equation? An equation is *a mathematical statement that has a left-hand side, an equal sign, and a right-hand side.* For example, $x + 5 = 7$ is an equation because it has a left-hand side, which is $x + 5$; an equal (=) sign; and a right-hand side, 7.

Second, to show how important equations are, we are going to name a few academic disciplines that use equations continuously: math, physics, chemistry, engineering, economics, research, statistics, and oceanography. This list is by no means complete; rather, it is an illustrative list of disciplines. More and more academic disciplines, such as geology and business, are now starting to use math modeling to predict new phenomena and/or to verify existing theories, and every one of them will use equations.

No matter what the discipline is, there is one thing that everyone does with an equation. *An equation is set up to solve for an unknown.* This unknown can be a variable and/or a quantity. For example, if $3x = 27$, then the solution of the equation involves finding the value of x.

What are the elements of this equation? Well, x is the unknown; 3 is the coefficient of x; and 27 is the value of $3x$. We want to know the value of x. In this particular case, the value of x is 9. How did we arrive at the answer algebraically?

Before we answer this question, we need to identify a very important part in this equation again. This very important part is the *coefficient* (the constants and/or variables around x) of the unknown quantity, x; the coefficient is 3 in this case. The method of solving this particular equation is to divide both sides of the equation by the coefficient of the unknown. That is, we divide both sides of the equation by 3 to get an answer of 9 for x, as shown:

$$3x = 27$$
$$\frac{3x}{3} = \frac{27}{3}$$
$$x = 9$$

Solving Linear Equations

The solution of linear equations (involving a single variable with a power of 1) involves the mastery of a variety of skills learned in earlier chapters and *pattern recognition*. Practice, practice, and more practice make it all easy; without mastery, this stuff will stay fuzzy in your mind, and the fuzziness will show up on a piece of paper as well. Here, we will discuss some common approaches that will be helpful to your students in the solutions of these equations.

Collecting and Transferring

The most basic and important technique in solving equations involves *grouping the unknowns on one side of the equal sign and the constants on the other side of it*. Note that when we transfer terms from one side of the equation to the other, *the terms change their signs*. Let us take some examples to illustrate these techniques.

Example 11.4. Solving for an Unknown in Simple Cases

	Problem	Explanation
1.	$x + 4 = 11$ $x = 11 - 4$ $x = 7$	You transfer 4, a constant, to the right side of the equation to be with 11, another constant. In doing so, you change the sign of 4. Therefore, it becomes -4 on the right side of the equation. When we subtract 4 from 11, we get 7 as the answer for x.
2.	$2x - 4 = 18 + x$ $2x - x = 18 + 4$ $x = 22$	Now, here is a little different twist. Here, we collect all of the x's on one side and all of the numbers on the other side. And in the process of this shifting around or transformation, we change the signs of those terms we have shifted. For example, x on the right hand side of the equal (=) sign becomes $-x$ on the left hand side. On the other hand, the constant -4 on the left side became +4 on the right side of the equation. Simplification gives an answer of 22 for x. Remind your students not to confuse coefficients with "stand-alone" constants.

Dividing by the Coefficient

This principle involves *dividing both sides of an equation by the coefficient of an unknown quantity*. Let us illustrate it with examples.

Example 11.5. Solving for an Unknown, continued ...

	Problem	Explanation
1.	$6x = 36$ $$\frac{6x}{6} = \frac{36}{6}$$ $x = 6$	Here the coefficient of x is 6. Therefore, we divide both sides by 6. On the left side, 6 cancels out. On the other hand, dividing 36 by 6 on the right side of the equation will result in an answer of 6.
2.	$5*6x = 120$ $30x = 120$ $$\frac{30x}{30} = \frac{120}{30}$$ $x = 4$	Here the coefficient of x is a multiplication of two numbers. Therefore, we simplify that multiplication first. The answer will be 30 if we multiply 5 times 6. Our last step will be to divide both sides of the equation by the coefficient of x, which is 30 in this case. This step results in answer of 4 for x.

Collecting, Transferring, and Dividing

Here we apply both principles as discussed earlier. To illustrate, here are some more examples.

Example 11.6. Solving for an Unknown with Multiple Terms

	Problem	Explanation
1.	$4x + 6 = 18$ $4x = 18 - 6$ $4x = 12$ $$\frac{4x}{4} = \frac{12}{4}$$ $x = 3$	Here the constants are collected on the right hand side of the equal sign. The constant +6 is transferred to the right side of the equation and its sign changes from + to -. Subtraction leads to remainder of 12. The last step is to divide both sides of the equation by the coefficient of x, which is 4. The final answer, therefore, is 3.
2.	$2x - 8 = -3x + 7$ $2x + 3x = 7 + 8$ $5x = 15$ $$\frac{5x}{5} = \frac{15}{5}$$ $x = 3$	In this case, the first task is to collect like terms by transferring all the unknowns to the left side and all the constants to the right side of the equation. Specifically, $-3x$ is transferred to the left and is changed to $+3x$. Similarly, -8 is transferred to the right and it changes to $+8$. The left-hand side becomes $5x$; the right hand side becomes 15. We divide each side by the coefficient of x, which is 5. The answer for $x = 3$ is obtained.

3. $5x + 7 - 7x = -4x + 13 - 1$ Again, collect all the like terms. Put all the x's on the left side and all constants on the other side, and be sure to change the signs of terms.

$5x - 7x + 4x = 13 - 1 - 7$

$2x = 6$

$\dfrac{5x}{5} = \dfrac{15}{5}$

$x = 3$

Then, simplifying each side gives $2x = 6$.

Now, we divide each side by the coefficient of x, which is 2. This results in answer of 3 for x.

Equations with Parentheses

Many equations that your students will come across involve parentheses. To solve them, the students must *first remove the parentheses by using the distributive law that calls for multiplication inside the parentheses* before proceeding any further with the equation.

Example 11.7. Solving for an Unknown with Terms in Parentheses

	Problem	Explanation
1.	$6x = 2(15 - 2x)$	Multiply the terms within the parentheses by 2, and remove them.
	$6x = 30 - 4x$	Transfer $-4x$ to the left side and change its sign.
	$6x + 4x = 30$	
	$10x = 30$	Simplify both sides
	$\dfrac{10x}{10} = \dfrac{30}{10}$	Divide both sides with the coefficient of x, which is 10.
	$x = 3$	
2.	$4 - 4(x - 4) = 2(x - 2) + 6$	Multiply within the parentheses on both sides and remove them. Be careful about the minus (-) sign outside the parentheses.
	$4 - 4x + 16 = 2x - 4 + 6$	
	$4 + 16 + 4 - 6 = 2x + 4x$	Collect unknowns on the right side and constants on the left side. In this case, we avoided the negative sign for x's by transferring x's to the right side of the equation.
	$18 = 6x$	
	$\dfrac{18}{6} = \dfrac{6x}{6}$	
	$3 = x$	Simplify terms on both sides. Divide both sides with the coefficient of x, which is 6.

SECOND CUMULATIVE REVIEW

Solve for the unknown.

1. $5x + 7 = 32$
2. $4y + 10 = 26$
3. $5x - 9 = 21$
4. $3y - 7 = 17$

5. $7x + 3 = -46$
6. $-45 = 4 + 7x$
7. $-4x + 6 = -18$
8. $-7x - 7 = -28$

Solve for the unknown.

9. $4x + 8x = 72$
10. $7x + 5x = 60$
11. $3x + 2x = 30$
12. $-4y - 5y = 45$
13. $-9y - 4y = -39$
14. $-4t - 3t = -56$

15. $8p - 30 = 3p$
16. $4n - 10 = -n$
17. $6y + 4 = 3y + 13$
18. $14 - 5r = -r + 2$
19. $4 - 2p = 3p - 7p + 24$
20. $4 + 5b - 7 = 4b + 2 - 2b + 7$

Solve for the unknown.

21. $3(2z - 3) = 21$
22. $6(3x + 2) = 30$
23. $2(3 + 4p) - 9 = 21$
24. $6 - 3(3b - 1) = 9$
25. $10 - 3(2x - 1) = 7$
26. $5r - (2r - 8) = 17$
27. $5(c + 4) = 6(c - 3)$
28. $4(f - 4) = 9(f - 4) - 5$
29. $9x + 5 - 7x = 19 - 12x + 10x - 10$
30. $5(w + 3) + 9 = 4(w - 2) + 6$

Word problems: Write an equation, and then solve for the unknown.

31. A number increased by 6 equals 9. What is the number?
32. The product of two numbers is 24. One of the numbers is 4. What is the other number?

33. The sum of two numbers is 19. One of the numbers is 11. What is the other number?
34. A number decreased by 6 equals 14. What is the number?
35. The difference of two numbers is 20. The smaller number is 16. What is the other number?

SAMPLE EXIT QUESTIONS

1. Solve for c: $5(c + 3) = 6(c - 2) + 2$

2. Solve for x: $k + x = 2x - 1$

3. A number decreased by 9 equals 11. What is the number?

SAMPLE PRIMING HOMEWORK

1. Solve for y: $\dfrac{y}{5} = \dfrac{20}{y}$

2. Solve for p: $\dfrac{3p}{100} = \dfrac{6}{2p}$

3. Solve for x: $\dfrac{2x}{y} = \dfrac{8z}{p}$

POINTS TO REMEMBER AND REVIEW

- Always simplify terms within parentheses first.
- In simplifying terms, three steps are required in this order: (1) the signs first, (2) followed by the coefficients of the variables, and (3) then the variables themselves.
- Use the distributive law for multiplying within parentheses. That is, multiply all of the terms *inside* the parentheses by the term and/or a number *immediately outside* the particular parentheses.
- Transpose terms as necessary to *collect all terms containing the unknown on one side of the equal sign and all the constants on the other side of it.* Terms *change their signs* when transposing them (bringing them across) to the other side of the equal sign.

12

The Algebra Savior

Cross-Products

In any work involving mathematical calculations at almost all levels, the *cross product*, sometimes known as *cross multiplication*, is one of the most important concepts used by students and professionals on a daily basis. In spite of the importance of mastering this concept, students at the high school and college freshman levels often have difficulty applying it. As a result, professors and high school teachers alike end up spending significant amounts of time explaining cross products to students again and again, even in some advanced courses.

For this reason, we have divided this chapter into three major sections. The first section deals with the concept of cross multiplication and then adds solving for an unknown after the task of cross multiplication is complete. The second section deals with applications involving cross multiplication that are simple but are used in almost all applications involving basic computation. The last section deals with problems involving somewhat more difficult cross multiplication.

REVIEW OF PREVIOUS KNOWLEDGE

In almost all science, technology, engineering, and mathematics (STEM) courses, the final step in problem solving or mathematical derivations is solving for an unknown. In its simplest form, the algebraic equation that involves solving for such an unknown can take the following form:

$$5x = 35$$

Here, the *unknown* is *x*, and the number 5 is the *coefficient* of *x*. One of the widely used techniques in solving for an unknown is to divide both sides of the equation by the coefficient of the variable. In this case, therefore, we divide both sides of the equation with 5. The answer for *x* will, therefore, be 7.

The mathematical steps for this process are:

Example 12.1. Solving for an Unknown in Simple Cases

Example	Explanation
$5x = 35$ $$\frac{5x}{5} = \frac{35}{5}$$ $x = 7$	Both sides are divided by the coefficient of *x*, that is 5. The division of 35 by 5 will give an answer of 7, and the left hand side will yield *x* by itself.

Although, this concept is a very basic, it is also a very important one in the STEM disciplines. Every student in the STEM disciplines and research-based courses must master this concept and the techniques it uses as soon as possible.

BASIC CONCEPTS IN CROSS MULTIPLICATION

The technique of cross products is used to solve a problem involving *proportion*. By definition, a *proportion* is an equation that states that *two ratios or fractions are equal*. An example of a proportion follows:

$$\frac{A}{B} = \frac{C}{D}$$

This proportion states that the ratio of A to B (A/B) is equal to the ratio of C to D (C/D). The numbers within the ratio on each side of the equal sign are said to be *proportional* to each other.

An example might look as follows:

$$\frac{n}{5} = \frac{6}{3}$$

Here the value of n must be such that the ratio $n/5$ is equal to the ratio $6/3$. In order to solve for the value of *n*, we use the concept of a cross product. In cross products or cross multiplication, *we multiply the elements that are diagonally across from each other*. Then we put an equal sign between the products. In this case, *n* is diagonally across from 3, and 5 is diagonally across from 6. Using cross multiplication, the process is:

$$\frac{n}{5} \bowtie \frac{6}{3}$$
$$n \times 3 = 5 \times 6$$
$$3n = 30$$

With the cross multiplication complete, now we have to solve for n. In order to solve for n, we must divide both sides of the equation by the coefficient of n, which is 3. Doing this division results in:

$$3n/3 = 30/3$$
$$n = 10$$

As a check, you can see that the ratio 10/5, which is 2, is also equal to the ratio 6/3, which is 2 as well.

SIMPLE CROSS PRODUCTS

We will illustrate this concept and its technique further by using somewhat different, but simple, examples.

Example 12.2. Solving for an Unknown using the Principles of Cross Products

	Proportion and solution	Explanation
1.	$\frac{n}{8} = \frac{5}{4}$ $4 \times n = 5 \times 8$ $4n = 40$ $4n/4 = 40/4$ $n = 10$	Multiply 4 and n, put an equal sign, and then place the product of 8 and 5 on the other side of the equation. Divide both sides of the equation by the coefficient of n, which is 4. This will result in an answer of 10 for n.
2.	$\frac{3x}{6} = 6$ $3x = 6 \times 6$ $3x = 36$ $3x/3 = 36/3$ $x = 12$	Before you perform the cross multiplication, you should mentally place 1 in the denominator under 6 on the right hand side of the equal sign, and then carry out the cross multiplication process. This will give you $3x = 36$. Now, divide both sides of the equation by the coefficient of x, which is 3. The value of x is 12.

Instruct your students to simplify *before* carrying out the actual multiplication. Also, remind them they need to write the product of two terms *as is* before carrying out the actual multiplication, because in most instances, these expressions will simplify to integers. For example, a cross product might look, as follows.

$$\frac{249}{249} = \frac{32}{16x}$$

If students multiply 249 by 32 and 249 by 16, either they need a calculator or it will take them quite a while to do the multiplication. An easy way out of this unnecessary dilemma is to simplify those terms that are on the left-hand side of the equal sign first and then simplify the terms on the right-hand side of the equal sign. In this example, 249/249 simplifies to 1. Similarly, 32/16 will simplify to 2. In the simplified form, the cross product can be written as:

$$\frac{1}{1} = \frac{2}{x}$$

Now the ratio terms are easy to cross multiply. The answer will be $x = 2$.

Exercise Set 12.1

Solve each proportion by cross multiplication.

1. $\dfrac{1}{2} = \dfrac{7}{m}$
2. $\dfrac{2}{3} = \dfrac{12}{b}$
3. $\dfrac{n}{8} = \dfrac{4}{2}$
4. $\dfrac{1}{3} = \dfrac{8}{m}$
5. $\dfrac{1}{3} = \dfrac{12}{m}$

6. $\dfrac{m}{2} = \dfrac{6}{2}$
7. $\dfrac{7}{4} = \dfrac{7}{p}$
8. $\dfrac{12}{5} = \dfrac{12}{q}$
9. $\dfrac{r}{8} = \dfrac{9}{8}$
10. $\dfrac{2}{2} = \dfrac{14}{x}$

11. $\dfrac{91}{5} = \dfrac{91}{y}$
12. $\dfrac{z}{28} = \dfrac{9}{28}$
13. $\dfrac{256}{2} = \dfrac{128}{r}$
14. $\dfrac{512}{4} = \dfrac{128}{s}$
15. $\dfrac{t}{8} = \dfrac{64}{32}$

ADVANCED CROSS PRODUCTS

Cross products can be quite cumbersome, but the underlying principle remains the same. In a simple case, as before, the numerators and denominators might contain simple variables (*x*, *p*, etc.) and/or constants (5, 11, etc.).

In some more complex situations, the numerators and/or denominators could contain multiple term expressions ($x + 11$, $t - 6$, etc.). We will illustrate these situations by offering here a number of examples.

Example 12.3. Solving for an Unknown using the Principles of Cross Products, continued …

	Problem	Explanation
1.	$\dfrac{n+1}{8} = \dfrac{5}{4}$ $4(n+1) = 5 \times 8$ $4n + 4 = 40$ $4n = 40 - 4$ $4n = 36$ $\dfrac{4n}{4} = \dfrac{36}{4}$ $n = 9$	Multiply 4 and $n+1$ after putting $n+1$ in parentheses to provide clarity to the operations, put an equal sign. Next, place the product of 8 and 5 on the other side of the equation. Multiply <u>both</u> terms within the parentheses by 4. Transfer the constant 4 from the left of the equal sign to the right of the equal sign, and change its sign. Simplify and then divide both sides of the equation by the coefficient of n, which is 4. This will result in an answer of 9 for n.
2.	$\dfrac{3}{9} = \dfrac{5}{2x - 5}$ $3(2x - 5) = 9 \times 5$ $6x - 15 = 45$ $6x = 45 + 15$ $6x = 60$ $\dfrac{6x}{6} = \dfrac{60}{6}$ $x = 10$	Multiply 3 and $2x - 5$, put an equal sign, and then place the product of 9 and 5 on the other side of the equation. You could have simplified the left side ratio, but it was not necessary because the numbers were small. Multiply both terms within the parentheses by 3. Transfer the constant -15 from the left of the equal sign to the right of the equal sign, and change its sign. Simplify and then divide both sides of the equation by the coefficient of x, which is 6. This division will result in an answer of 10 for x.

Exercise Set 12.2

Solve each proportion by cross multiplication.

Instruct your students to simplify before carrying out the actual multiplication. Also, remind them they need to write the product of two terms before carrying out the actual multiplication because in most instances these expressions will simplify to single-digit integers. Also, ask your students to slow down, and take … their … time. If they don't, they can make some silly mistakes.

1. $\dfrac{1}{2} = \dfrac{7}{m+1}$
2. $\dfrac{2}{3} = \dfrac{12}{b+2}$
3. $\dfrac{n+1}{8} = \dfrac{4}{2}$
4. $\dfrac{1}{3} = \dfrac{8}{m+3}$
5. $\dfrac{1}{3} = \dfrac{12}{m+4}$
6. $\dfrac{m+2}{2} = \dfrac{6}{2}$
7. $\dfrac{7}{4} = \dfrac{7}{p+5}$
8. $\dfrac{12}{5} = \dfrac{12}{4q+1}$
9. $\dfrac{6r+9}{8} = \dfrac{9}{8}$

CHALLENGING PROBLEMS

Simplify *before* doing cross multiplication.
   Instruct your students to take the time to study these problems. They need to recognize patterns before jumping into solving these kinds of problems.

1. $\dfrac{0.123}{0.123} = \dfrac{2x}{10}$
2. $\dfrac{0.25}{0.50} = \dfrac{2x}{16}$
3. $\dfrac{5000}{500} = \dfrac{2x}{10}$
4. $\dfrac{6}{x} = \dfrac{x}{6}$
5. $\dfrac{x}{4} = \dfrac{16}{x}$
6. $\dfrac{20}{x} = \dfrac{2x}{10}$
7. $\dfrac{0.25}{0.125} = \dfrac{2x}{10}$
8. $\dfrac{100}{1000} = \dfrac{2x}{20}$
9. $\dfrac{256}{128} = \dfrac{2x}{10}$
10. $\dfrac{20}{40} = \dfrac{2x+2}{12}$
11. $\dfrac{10}{20} = \dfrac{2x-3}{10}$
12. $\dfrac{0.10}{0.01} = \dfrac{2x-4}{2}$

MENTAL CHAPTER REVIEW

Your students need to do these problems mentally in the class. Guide this effort to ensure that all of the students are engaged in answering these questions. Guarantee this participation by calling upon students by name and by making use of their differential knowledge bases. Prepare a handout of these problems to save time, or simply write them on the board.

1. $\dfrac{1}{3} = \dfrac{7}{m}$
2. $\dfrac{3}{3} = \dfrac{12}{b}$
3. $\dfrac{n}{8} = \dfrac{4}{8}$
4. $\dfrac{1}{4} = \dfrac{8}{m}$
5. $\dfrac{1}{12} = \dfrac{12}{m}$
6. $\dfrac{m}{3} = \dfrac{6}{2}$
7. $\dfrac{17}{4} = \dfrac{17}{p}$
8. $\dfrac{128}{5} = \dfrac{128}{q}$
9. $\dfrac{r}{8} = \dfrac{34}{8}$
10. $\dfrac{2}{2} = \dfrac{81}{x}$
11. $\dfrac{91}{13} = \dfrac{91}{y}$
12. $\dfrac{z}{281} = \dfrac{9}{281}$
13. $\dfrac{64}{2} = \dfrac{128}{r}$
14. $\dfrac{64}{4} = \dfrac{128}{s}$
15. $\dfrac{2t}{2} = \dfrac{32}{32}$

SAMPLE EXIT QUESTIONS

Solve for the unknown.

1. $\dfrac{32}{4} = \dfrac{16}{s}$
2. $\dfrac{27}{x} = \dfrac{x}{3}$
3. $\dfrac{m-5}{121} = \dfrac{-5}{121}$

SAMPLE PRIMING HOMEWORK

1. If 20 lb of turkey breast contains enough meat for 60 servings, how many pounds of turkey breast will be needed for 300 servings?
2. How tall is a flagpole that casts an 88-ft shadow at the same time that a 5-ft lady casts an 11-ft shadow?

POINTS TO REMEMBER AND REVIEW

- You can simplify terms in the numerators and denominators on the same side of the equal sign.
- You can also simplify terms in the numerators and denominators on both sides of the equal sign.
- However, you *cannot* simplify terms that are diagonally opposite of each other —these terms can only be multiplied.
- Simplify before you multiply.
- If, after cross multiplication, the equation ends up with the square of a variable, there is one more step that needs to be completed: *take the square root of both sides* before you get a final answer.
- Use the distributive property when multiplying a numerator or denominator that contains more than one term (variable or constant).

13

Ratio and Proportion

While ratio and proportion continue to be a big issue, this topic is very easy to teach and understand when the problems are divided appropriately into their parts. In their simplest form—which seems to be the case most of the time—ratio and proportion problems involve a total of four terms: two sets of them, each set containing two of the same type. Three of the terms (variables) are usually given, and students are asked to solve for the fourth one (the unknown). In addition, cross products (cross multiplication) can always be used to solve for the unknown term.

Before we discuss this topic further, review some previously learned knowledge with the students. In this instance, the needed knowledge consists of solving a simple equation for an unknown *after* performing the operation of cross products.

REVIEW OF PREVIOUS KNOWLEDGE

Solving for an Unknown in a Simple Equation

Example 13.1. Solving for an Unknown in Simple Cases

Example	Explanation
$4x = 24$ $$\frac{4x}{4} = \frac{24}{4}$$ $x = 6$	Divide both sides of the equation by 4, the coefficient of x. This division will give the answer of 4 for x.

> General Instruction for the teacher:
> Either prepare a handout for the problems in the sections marked as exercise sets, mental problems, and review exercises or write these problems out on the board before seeking answers from your students. Again, use the principle of differential knowledge base to engage students in their own learning, and always call them by their first names. Make certain to ask students to wait to ask a question or respond to a question until recognized, or the "swifties" will take over.

Exercise Set 13.1

Solve for x.

1. $5x = 25$
2. $10x = 100$
3. $6x = 36$
4. $20x = 100$
5. $15x = 45$
6. $54 = 9x$
7. $9x = 81$
8. $40 = 4x$
9. $50 = 10x$

Cross Products

Let us use an example first.

Example 13.2. Solving for an Unknown using the Principles of Cross Products

Example	Explanation
$\frac{6}{3} = \frac{x}{7}$ $3x = 42$ $\frac{3x}{3} = \frac{42}{3}$ $x = 14$	In this case, you multiply 6 and 7, and then 3 and x. After this procedure, equate the two products. Then, divide both sides of the equation by 3, the coefficient of x, to obtain $x = 14$

Exercise Set 13.2

Solve for x.

1. $\frac{5}{2} = \frac{x}{6}$
2. $\frac{9}{3} = \frac{x}{7}$
3. $\frac{10}{4} = \frac{x}{6}$
4. $\frac{6}{3} = \frac{x}{15}$
5. $\frac{12}{8} = \frac{x}{6}$
6. $\frac{9}{3} = \frac{2x}{10}$

PROCEDURES FOR SETTING UP AND SOLVING PROBLEMS

In ratio and proportion, your students will come across two ratios. A ratio involves two quantities. A proportion deals with two ratios, which means two sets of quantities—four quantities in total. For example, a proportion can contain the two heights and two weights (four quantities) of two different people (two ratios).

Problem Setting in Ratios

A ratio is a quotient of two quantities: one in the numerator and the other in the denominator. For example, the word problem "the ratio of 6 to 8" can be written as $\frac{6}{8}$. This expression is also known as the *fractional notation* for this ratio. The message here is that *the quantity that follows the word "to" in a word problem is written as the denominator.*

Exercise Set 13.3

Find the ratio and simplify mentally.

Avoid writing these problems on the board. The more easily your students can do these problems mentally, just by listening to you, the better off they will be.

1. 4 to 6
2. 6 to 10
3. 9 to 12
4. 18 to 24
5. 20 to 100
6. 100 to 1000

Setting up and Solving Proportion Problems

Setting up a particular problem is usually quite easy. Every quantity in the problem will have units—weight in pounds, height in inches, time in seconds, and so forth. The four quantities in the ratio belong to two sets. The quantities of the first are written in the numerator and denominator. Place an equal sign followed by the quantities of the second set in the numerator and denominator. You must ensure that *the units of the quantities in the numerator are the same* (e.g., pounds and pounds) and *the units of the quantities in the denominator are the same* (e.g., feet and feet).

Let us illustrate this principle with an example: Linda has a fuel-efficient car. On a full tank of 10 gallons of gas, she can drive 400 miles. How many miles can she drive if her car's tank has only 5 gallons of gasoline?

An accurate analysis of this problem shows that there are two sets of quantities here. The first set consists of 10 gallons and 400 miles; the second set consists of 5 gallons and an unknown number of miles, which is what we want to solve for. Keep in mind that the numerators of both sides must be in the same units, either gallons or miles. We will arbitrarily choose miles for the numerator and gallons for the denominator for both sets of quantities. We then write them down as follows, separated by an equal sign:

$$\frac{400 \text{ miles}}{10 \text{ gallons}} = \frac{y \text{ miles}}{5 \text{ gallons}}.$$

Before you take the cross product, you can also write this proportion as one without the units to avoid confusion.

$$\frac{400}{10} = \frac{y}{5}$$

Simplifying will give

$$\frac{40}{1} = \frac{y}{5}.$$

The cross product or cross multiplication gives you

$$1 \times y = 40 \times 5$$

or

$$y = 200 \text{ miles},$$

which is the answer to the problem.

You can apply this principle to all problems involving ratio and proportion.

EQUAL RATIOS

Two ratios are equal if their cross products are equal. Consider an example:

$$\frac{8}{12} = \frac{2}{3}.$$

These two ratios are equal because their cross products (8×3 and 12×2) are both equal to 24. Therefore, these two ratios are equal. *We define a proportion as two ratios that are equal.*

REVIEW PROBLEMS

Encourage your students to complete mentally as many problems as possible.

Solve for the unknown variable.

1. $6x = 36$
2. $9x = 72$
3. $15x = 60$
4. $40 = 8x$
5. $55 = 5x$
6. $\dfrac{11}{3} = \dfrac{x}{6}$
7. $\dfrac{12}{8} = \dfrac{3x}{6}$
8. $\dfrac{9}{5} = \dfrac{2x}{10}$

Find the cross products for each of the problems to determine if the following ratios are equal or not.

9. $\dfrac{10}{3} = \dfrac{20}{6}$
10. $\dfrac{12}{3} = \dfrac{20}{7}$
11. $\dfrac{15}{10} = \dfrac{9}{6}$

Set up as a ratio of two terms before answering the questions.

12. A race car driver drives around a 5-mile track in 5 minutes. How long will it take to drive 20 miles?
13. In 2 minutes, Jose can skate 1500 meters. How far can Jose skate in 6 minutes?
14. It takes Mitchell 200 seconds to swim 300 meters. How far would you expect Mitchell to swim in 50 seconds?
15. John is 6 feet tall and weighs 180 pounds. What do you think will be the approximate weight of Peter who is 5 feet tall?

Find the ratio and simplify.

16. 2 to 3
17. 4 to 8
18. 8 to 16
19. 4 to 16
20. 50 to 100
21. 100 to 1000

CHALLENGING PROBLEMS

Do half of these problems in the class, engaging all the students. Take this opportunity to emphasize the writing of similar terms (quantities with the same units) in the numerator and the denominator.

1. A television screen has a width of 48 inches and a height of 32 inches. What is the ratio of width to height? What is the ratio of height to width?

2. The Leaning Tower of Pisa is approximately 180 ft tall, but leans about 20 ft from its base. What is the ratio of the distance it leans to its height? What is the ratio of its height to the distance it leans? Simplify your answers.
3. A quality-control inspector examined 200 light bulbs and found 20 of them to be defective. At this rate, how many defective bulbs will there be in a lot of 4000?
4. Rachel can waterproof 500 ft² of her decking with 3 gallons of sealant. How many gallons should Rachael buy for a 1500 ft² deck?
5. If 10 pounds of turkey breast contains enough meat for 40 servings. How many pounds of turkey breast will be needed for 200 servings?
6. In 2005, 1 U.S. dollar was worth about 100 Japanese yen. How much was $10 worth in yen? Jeffrey, when traveling in Japan, bought a fancy toy for his kids that cost him 8000 yen. How much did it cost him in dollars?
7. How tall is a flagpole that casts a 77-ft shadow at the same time that a 5-ft person casts a 7-ft shadow?
8. The two scalene triangles below are similar. Find the missing length x.

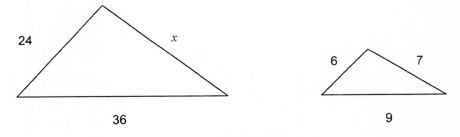

SAMPLE EXIT QUESTIONS

Solve for the unknown.

1. $\dfrac{3y}{5} = \dfrac{12}{5}$
2. $\dfrac{4x}{80} = \dfrac{5}{x}$
3. A quality-control inspector examined 500 light bulbs and found 50 of them to be defective. At this rate, how many defective bulbs will there be in a lot of 10000?

Ratio and Proportion

SAMPLE PRIMING HOMEWORK

1. The price of shoes that you liked in a store was originally $200. The shoes went on sale and were marked "20% off." What is the *discount*, and what is their *sale price* after the discount?
2. A friend of yours works as a salesman at a car dealership. He receives a commission of 5% on every car he sells. During the month, he sold three cars worth a total of $60,000. How much commission did he make during this month?
3. What is the simple interest on $500 invested at an interest rate of 4% for one year?

POINTS TO REMEMBER AND REVIEW

- A proportion deals with four quantities, but only of two different types—for example, two distances and two quantities of gasoline. A sequence of four steps is required to make sure the solution is accurate:

  ✓ First, set the problem mechanically: draw a horizontal line, put an equal sign, and draw another horizontal line.
  ✓ Next, in the numerator of both, write the units for one quantity—for example, *miles* for distance—and then in the denominator write the other quantity's units, in this case, *gallons* (eventually, you should be able to do this process mentally). Then actually write the numbers for one set on the left side of the equal sign, and the numbers for the second set to the right of the equal sign, using a variable like x for the unknown quantity, for example, $\dfrac{400 \text{ miles}}{20 \text{ gallons}} = \dfrac{1000 \text{ miles}}{x \text{ gallons}}$.
  ✓ Then simplify first before you take the cross product (try to cancel zeros in the numerator and denominators of the left and right side of the equal sign—but don't cancel zeros diagonally; you only *multiply* in the diagonal direction.
  ✓ Finally, solve for the unknown.

14

Just Shopping for the Best Deal

Discount, Sales Tax, Commission, Percentage, and Simple Interest

The five concepts of discount, sales tax, commission, percentage, and simple interest are related, and they usually are used when dealing with money and/or shopping. In any shopping situation, an item may be marked down for clearance at a certain discount or percentage of the original price, and someone in the store may make a commission on the item when it is sold. Then when you check out, you usually pay a sales tax, a percentage. The same principle applies in investing money. For example, people look for a bank that will provide a higher return (percentage) on a deposit than will another bank.

Let's look at an example. Lilly goes to a superstore and sees a dress that was originally priced at $100, but now it is on sale. It is marked as being "45% off." How much will Lilly pay for this dress? We are sure your students will get the answer of $55 as the *discounted price* (also known as the *sale price*) for the dress, without ever setting it up as a problem. The question is how did they get the right answer?

The students will probably have figured out that 45% of $100 is $45, which is the discount. If they subtract $45 from $100, they will get an answer of $55. The first part of this question can also be written as, "What is 45% of $100?" The second part is subtracting this answer from the original price of $100 to get the sale price. So, these two concepts are related.

Whereas a discount is used to lower the price of an item, a commission is used to boost a person's salary. In contrast, simple and/or compound interest rates will be used by customers to increase their financial principal by buying certificates of deposit or investing in some other interest-bearing type of account. Credit card companies apply the concept of interest rate to increase their profits by letting you continue to charge on your credit cards without paying the full amount of your purchases every month, while charging you interest on the unpaid balance. (This fact could be a good wake-up call for your students who may want to examine their own credit card payments a little more carefully to see how much they are really paying.)

The question is: How does it all work? How do we compute these various quantities when the numbers are complex and cannot be easily simplified mentally? The answer is not that straightforward, but students can make it easier for themselves by becoming good at the process of cross products and solving for an unknown in simple situations. The rest will follow.

Begin with a review of cross products (chapter 12) and solving for an unknown in simple situations. These two processes are critical: without them, your students will experience difficulties in this chapter. Also, let your students know—and keep on emphasizing to them—that they should use the golden mathematical rule of *simplify first* before carrying out tediously involved multiplications or divisions. Most of the time, students will be able to do that.

So, let us look at discounts.

FINDING A DISCOUNT

Discount problems are always the same. Items are marked down by a certain percent from the original price. Then what the consumer has to do is to find the discount amount and the sale price. Let us illustrate this task by an example: Maggie wants to a buy a blouse that was marked 30% off the original price of $50. Two questions are usually asked. What is the discount? What is the sale price?

This is a typical problem when dealing with discounts. Look carefully at this problem. We could have asked it as: What is the actual discount on a blouse that is marked down 30% from the original price of $50? There is a way of solving such a problem. Your students need to commit to memory the following rules, once and for all, before proceeding further with the solutions of such problems.

In discount problems, the *discount rate* and *discount amount* are two different quantities. The discount *rate* is always shown *as a percentage*. The discount *amount* is always a *number*, usually a dollar amount in the United States. The actual sale price is always the *difference* between the original price and the discount. Mathematically, these quantities can be written as a formula for discount as:

$$\text{Discount amount} = \frac{\textit{Discount rate in \%}}{100} \times \text{Original amount}$$

$$\text{Sale price} = \text{Original price} - \text{Discount amount}$$

RULES FOR TRANSLATING WORDS INTO MATHEMATICAL OPERATIONS

Rules for translating key words in a problem into mathematical operators

Word	Translation
what	unknown quantity (x, y, z, p, q, m, n, etc.)
of	multiplication (*, x)
is	equal (=)
n% notation	change to fractional notation (divide n% by 100); also known as decimal notation

Box 14.1 Rules for Translating Words into Mathematical Operations for Percent Problems

Let us write the problem one more time to illustrate these rules. Using these rules in the above problem will lead to:

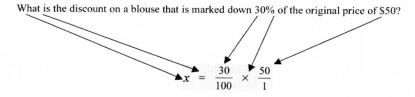

What is the discount on a blouse that is marked down 30% of the original price of $50?

$$x = \frac{30}{100} \times \frac{50}{1}$$

Notice, in this setup, you will see a 1 in the denominator beneath 50. Dividing by 1 does not change the answer, but it clarifies all the numerators and denominators.

Now simplify before you multiply. What do we mean by this? We mean don't multiply 30 and 50 first and divide by 100. Instead, cancel two zeros in the denominator with two zeros (one with 30 and the other one with 50) in the numerator. We can do this because 30 = 10 * 3 and 50 = 10 * 5. Then if you multiply 10 by 10, it will give you 100 in the numerator, which will cancel out with 100 in the denominator. Phew!

It takes a lot of words to explain the process, but it is rather simple, once the students understand it and get in the habit of doing it. Let us show you this process by putting it in an equation form.

$$x = \frac{30}{100} * \frac{50}{1} = \frac{(3 \cdot 10)}{100} * \frac{(5 \cdot 10)}{1} = \frac{3 \cdot 5 \cdot 10 \cdot 10}{100} = \frac{3 \cdot 5 \cdot 100}{100} = 15$$

Essentially, what we have done with this equation is this:

$$x = \frac{30}{100} * \frac{50}{1} = \frac{3 \cdot 5}{1}$$

Here, we have canceled the two zeros in the numerator with two zeros in the denominator. *But, please be careful when you do that.* So, the answer is $x = 3 * 5 = 15$, meaning the *discount amount* is $15.

To find the answer for sale price, we use the formula

Sale price = Original price – Discount amount,

which means the sale price is 50 – 15 = $35.

There is no need to put the $ sign everywhere, because it takes too long and distracts from the actual mathematical process. However, be sure to put the dollar sign at the end of the solution as shown here, so that the problem is completely answered.

Exercise Set 14.1

Instruct your students to practice writing down all the formulas they use when doing the problems. Shortcuts are not recommended for such problems.

1. Jeff's school's bookstore had an algebra book with an original price of $200 marked as 40% off the original price. What is the discount? What is the sale price?
2. Your neighborhood grocery store had a 25% discount sale on all cookies. Five boxes of chocolate cookies were originally priced at $10.00 in total. What is the discount? What is the sale price?
3. Your local electronics store had a flat-screen TV originally costing $2500 on sale. The sale price was 20% off the original price. What is the discount? What is the sale price?
4. Christina bought a fax machine for her dad as a Christmas gift. The original price was $300, but it was on sale and was marked as 30% off. What is the discount? What is the sale price?

SALES TAX

Ask the students, "Have you had the experience of buying a calculator listed at $10.00, and when you try to pay at the cash register, the cashier tells you the charges are not $10.00? You find out the full price is more like $10.50 or maybe $10.75. You ask why?"

In a country like ours, the federal government has bills to pay, states have payrolls to meet, and counties and cities need to pay their employees and take care of the essential neighborhood services. For all these activities and services, federal, state, and local governments need money. Where do they get the money to be able to pay? From all of us.

Just Shopping for the Best Deal

The money they collect is in the form of taxes. One of the most common forms of taxation is called a *sales tax* and is shown as a percent. Let us illustrate this by an example. Let's say you go to McDonald's and buy a sandwich, Coke, fries, and some cookies. You add all this up in your head, and you calculate it should cost $8.00. However, you receive a surprise: the bill is more than $8.00. If you are lucky and the sales tax in your state is a low 5%, how much do you have to pay at the cash register?

Formula for Sales Tax

Once again, as with the case of discounts, we are dealing with three variables: the sales tax *amount*, sales tax *rate*, and the *purchase price*. The formula for the sales tax amount is very similar to that for the discount amount:

$$\text{Sales tax amount} = \frac{\textit{Sales tax rate in \%}}{100} \times \text{Original price}$$

To obtain the total cost to you—the total purchase price or the total you must pay at the cash register—you simply add the sales tax amount to the initial purchase and/or sticker price. The formula for the total price is:

$$\text{Total price} = \text{Original price} + \text{Sales tax amount}$$

The problem regarding your purchase at McDonald's can also be stated as:

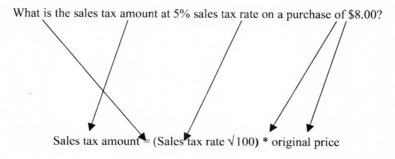

What is the sales tax amount at 5% sales tax rate on a purchase of $8.00?

Sales tax amount = (Sales tax rate √100) * original price

$$\text{Sales tax amount} = \frac{5}{100} * \frac{8}{1} = \frac{40}{100} = \$0.40 = 40 \text{ cents}$$
$$\text{Total price and/or cost} = \$8 + \$0.40 = \$8.40$$

Don't forget to change the percent into fraction (decimal) notation by dividing it by 100.

Exercise Set 14.2

1. Bill buys an old car for $2000 in Maryland, where the sales tax rate is 6%. What is the amount of the sales tax? What is the total purchase price, including the original price and tax?
2. Roger went to a computer show in Las Vegas where he saw a beautiful laptop for $900. The sales tax in Las Vegas is 7%. What was the amount of the sales tax he paid? What was the total Roger paid for the laptop?
3. You went to a store to help your parents buy a washer and dryer in New York. Both of them were priced at $700 each. The sales tax in New York is 8%. What was the sales tax amount? What was the total purchase price that your parents paid at the cash register?
4. Cathy and her mother purchased a DVD player that was on sale for $200 in Virginia, where the sales tax is 5%. What was the sales tax? How much did they pay at the cash register?
5. The tax rate in New Jersey is 6%. Brian bought a pair of shoes and some socks. The shoes were kind of expensive and cost him $135. The cost of the socks was $15.00. What was the sales tax? What was the total purchase price he paid at the cash register?

COMMISSIONS

A lot of us work for a salary, and our monthly check is the same until we receive a pay raise. Then there are those men and women who work on commissions only, and still others who work on a combination of the two. We are primarily concerned here with those workers who are paid a percent of the sales they make, which is called a *commission*.

Formula for Computing Commissions

Again, the principle is very simple. As before, three variables are involved with a problem of this nature: the *commission amount*, the *commission rate*, and the *total sales*. Mathematically, the equation for computing the commission amount is:

$$\text{Commission amount} = \frac{\text{Commission rate in \%}}{100} \times \text{Total sales price}$$

Let's look at an example to illustrate the use of this equation. Peter works on commission in car sales. His commission is 4% of his total sales for the month. He sells cars worth a total of $100,000 in one month. We want to know what Peter's commission is for the month.

To solve this problem, we use the equation for commission amount.

$$\text{Commission amount} = \frac{4}{100} * \frac{100000}{1} = 4 \times 1000$$

After simplification, we find that his commission is $4000.

Keep in mind that your students *always need to change percents to fraction and/or decimal notations before they can use them in any equation*, and they do it by *dividing the percent rate by 100*.

Exercise Set 14.3

1. A real estate company earns 6% of a sale in commission on home sales. If an agent sells a house for $600,000, what is the commission amount?
2. An agent working for a furniture company could earn 8% on all of her sales. If the agent sold $50,000 worth of merchandise, what is the agent's commission?
3. Pat works in a department store in the young women's department. She earns a commission of 11% on every item that she sells, plus her minimal basic pay. She sold only $3000 worth of merchandise during the slow month of January. What was Pat's commission for the month?

PERCENTAGES

This area is perceived to be the most difficult by many of our students. First, a big problem is actually setting up the problem itself, but the more difficult task is what to do once the problem is set up. A percent problem involves cross products and solving for an unknown. A further difficulty is that problems involving percents also can become quite cumbersome because students have a tendency to multiply everything before they simplify.

The good news, however, is that the problems in percents do have a logic and some rules to follow. However, students must master these rules, and that takes practice, practice, and more practice. With enough practice, this topic will become easy for them in due course. The key here is to provide plenty of practice with practical, real-life problems.

So, what's involved here? The problems in percentages involve three variables—the percentage, the base, and an amount—in an equation. Two of these are always given; students have to learn how to set up an equation to solve for the third one. Here is a typical problem in percentages:

30% of what number is 60?

Before solving this problem, let us recall the rules of translating some of the key words in this problem to mathematical operators.

Rules for Translating Percentage Problems

What is translated to a variable, like *n*, that is to be found; *is* means an equal sign; *of* goes into a product sign (×); and % is changed to a fractional notation with 100 in the denominator before using it in an equation. Let us follow the rules by restating the question above into an equation format. We have added some extra spaces between words for clarity purposes.

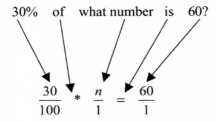

In this problem, we have used the property of 1—if you divide by 1 or multiply by 1, it does not change the value. Also, using this approach *clearly identifies the numerators and denominators*. Next, we can simplify the left side as $\frac{3n}{10}$ because 10 divides evenly into both 30 and 100. The equation, therefore, simplifies to:

$$\frac{3n}{10} = \frac{60}{1}$$

Cross multiplying at this step will lead to

$$3n = 60 * 10$$
$$3n = 600$$

Dividing both sides by 3, the coefficient of *n*, will lead to

$$\frac{3n}{3} = \frac{600}{3}$$

The final answer, therefore, is

$$n = 200$$

Exercise Set 14.4

1. What is 11% of 200?
2. What is 3% of 50?
3. What is 1% of 10?
4. 20% of what is 200?
5. 5% of what is 50?
6. 30 is what percent of 300?
7. 5 is what percent of 500?

SIMPLE INTEREST

Billy, who is 18 years old, has saved a couple of thousand dollars during his teen years. One day, when he was in a conservative mood, he asked his father, William, how he could invest his money safely. His father suggested there are at least two different ways that are safe. Their return on the investments may be slightly lower in comparison with other types of investments, but such investments are quite safe. The two safe methods his father suggested were government bonds and certificates of deposit (CDs). Billy decided to invest his money in CDs. (Banks usually pay interest on these CDs that is compounded daily, monthly, or yearly. To simplify things a little bit, we will only deal with simple interest because the equation is easier to use.)

The equation for simple interest is:

$$\text{Interest amount} = \text{Principal} * \frac{\text{Interest rate in \%}}{100} * \text{time in years}$$

The *principal* is the original amount invested. Symbolically, we write this equation as:

$$I = \frac{P*r*t}{100},$$

where:

I = interest earned
P = principal amount
r = rate of interest in %
t = time in years.

The important point to remember when using this equation is that the time must always be in years. If the time is in months, then change it to years by dividing it by 12, because there are 12 months in a year. Six months will translate into 6/12 (= 1/2) for t. If the time is three months, it will translate into 3/12 (= 1/4).

Let us use an example to illustrate this. What is the simple interest on $1200 invested at an interest rate of 4% for one month? Using the interest formula, we get:

$$\text{Interest} = \frac{1200 * 4 * 1}{100 * 12} = 4.$$

The interest will be $4.00.

Exercise Set 14.5

We suggest you do one or two problems in class to create self-confidence in your students and also to minimize frustration on everybody's part.
1. What is the simple interest on $3000 invested at an interest rate of 7% for one year?
2. What is the simple interest on $2000 invested at an interest rate of 6% for six months?
3. What is the simple interest on $4000 invested at an interest rate of 5% for ninety days (three months)?
4. What is the simple interest on $1200 invested at an interest rate of 4% for one month?
5. What is the simple interest on $5000 invested at an interest rate of 8% for three years?

CHALLENGING PROBLEMS

You should try to do the even-numbered problems in class, using the differential knowledge bases of your students. Then have your students complete the rest of the problems either at home or right after the class if their schedules permit it.

1. Jeff bought a pair of brand name tennis shoes for $80 that were originally priced at $120. What was the discount amount? What was the discount rate? *Hint:* Use the equation:

    Discount amount = Discount rate * Original price

2. Joe and his wife bought a TV on sale for $480 that was marked as being 40% off the original price. What was the original price of the TV? *Hint:* Use the equation:

    $$\text{Original Price} = \frac{\textit{Sale price}}{[1 - (\textit{Discount in \%}/100)]}$$

3. The Smith family paid $10 in sales tax in a restaurant for food and drinks. The bill before the tip was $200. What was the sales tax rate?

4. The Smiths decided to give a $40 tip for the excellent service they received on a food bill of $200, without the sales tax. What percentage of the pretax bill was the tip?
5. Bonnie earned a commission of $200 selling $2000 worth of merchandise. What was her commission rate?
6. Joe earned a commission of $1000 selling $20,000 worth of carpets. What was his commission rate?
7. What is 30% of $30?
8. What percentage of 300 is 30?
9. Trish invested her $200 in savings from a Christmas bonus in a certificate of deposit at 5% simple interest for three years. How much interest did she earn during this time? What is the total amount of money that the bank must pay after this period?
10. Sally invested $2000 in bonds for 3 months at 4% simple interest. What were her earnings from this investment after this period?

MENTAL CHAPTER REVIEW

Again, do the even-numbered problems in class, using the differential knowledge bases of your students. Your students should complete the rest of the problems during the class.

1. Steve's school's bookstore had a geometry book with an original price of $200 marked as 30% off the original price. What is the discount? What is the sale price?
2. Your neighborhood grocery store had a 35% off sale on all cookies. Five boxes of chocolate cookies were originally priced at $20.00. What is the discount? What is the sale price?
3. Paul buys an old car for $3000 in Maryland where the sales tax rate is 6%. What is the amount of the sales tax? What is the total price, including the original cost plus tax?
4. Peter went to a computer show in Las Vegas where he saw a beautiful laptop for $1200. The sales tax in Las Vegas was 8%. What was the amount of the sales tax? How much in total did Peter pay for the laptop?
5. A real estate company that earns 6% in commission on home sales sells a house for $500,000. What is the commission amount?
6. An agent working for a furniture company earns 10% on all of the sales she makes. If this agent sold $60,000 worth of merchandise, what is her commission?
7. What is 12% of 300?
8. What is 5% of 60?
9. What is 1% of 20?

10. 20% of what is 400?
11. 6% of what is 60?
12. The number 30 is what percent of 600?
13. The number 7 is what percent of 700?
14. What is the simple interest on $4000 invested at an interest rate of 8% for one year?
15. What is the simple interest on $3000 invested at an interest rate of 5% for six months?

SAMPLE EXIT QUESTIONS

1. The sum of two even consecutive numbers is 26. What are the two numbers?
2. What is the simple interest on $4000 invested at an interest rate of 5% for three months?
3. 9% of what is 90?

SAMPLE PRIMING HOMEWORK

1. The diameter of a circle is 20 cm. What is its radius?
2. If the radius of a circle, r, is 5, what is r^2?
3. If the radius of a circle, r, is 4, what is r^3?
4. Solve for L: $40 = 2(L + 3L)$
5. Solve for L: $100 = (L * 4L)$

POINTS TO REMEMBER AND REVIEW

- Always set up a problem according to the rules. In percentage problems:
    - ✓ *is* translates to an equal sign;
    - ✓ *of* translates to a multiplication sign,
    - ✓ a *% sign* means that you are to divide by 100, and
    - ✓ the word *what* translates to an unknown quantity, x.

- Simplify first before multiplying or dividing.
- In a discount problem, calculate the discount first. The final price will be less than the original price—you *subtract* the discount amount from the original price.
- In a tax problem, calculate the tax first. The final cost will be more than the price of the item—you *add* the tax to the original cost.
- An expression like "5 less than x" translates into $x - 5$, *not* $5 - x$.

15

How Big?

Perimeter, Area, and Volume

Let us consider some questions about the size of objects. For example, how big is the Earth? How big is your home? How large is our classroom? When we ask these questions, we are also asking the related questions: How long is it? How wide is it? What is the area? What is the volume? This chapter teaches a way of computing the perimeters of angular shapes, also known as the circumference in the case of a circle, and the surface areas of particular shapes.

Also, this chapter shows how to find the volume of some regular shapes. So, here we deal with objects in two and three dimensions. The familiar shapes we have in mind are the squares, rectangles, triangles, and circles as the 2-D objects and solids such as cubes, spheres, cylinders, and cones as 3-D objects.

The bottom line is that it does not matter what form a particular shape takes. There always are mathematical formulas that will give your students the information needed to solve a problem regarding a given shape. All your students have to do is identify these quantities in a given problem and then solve for the unknown. Once your students learn the art of problem solving in these areas, it becomes a piece of cake to solve scary-looking problems dealing with area and volume. Thus, the emphasis of the problems in this chapter is on the techniques of solving problems of area and volume rather than on the drawing of hundreds of shapes.

REVIEW OF PREVIOUS KNOWLEDGE

We begin this chapter with a review for your students as to their knowledge of solving for an unknown in a simple equation involving squares, cubes, square roots, and cube roots. The examples presented next are to assist you with this review. Be sure to involve all of the students in this review—not just the outspoken ones in the front seats. Also, do not, under any circumstances, write the information on the chalkboard and consider that a review. The information for the review must come from *students* and *not* from *you* as the teacher.

Example 15.1. Review of Solving for an Unknown, Cubes and Cube Roots

	Example	Explanation
1.	$2a \times 5 = 40$ $10a = 40$ $\dfrac{10a}{10} = \dfrac{40}{10} =$ $a = 4$	Multiply all constants around the unknown quantity. In this case the unknown quantity is a; 2 and 5 are factors around it called coefficients. Divide both sides of the equation with the coefficient of a, in this case 10. The answer for a is 4.
2.	$(2)^3$ $= 8$	Here 2 is raised to the power of 3, You want to see three 2's in multiplication; $(2)(2)(2)$ The answer is 8.
3.	$\sqrt[3]{27}$ $= 3$	Here we need to find a number when multiplied by itself three times, gives an answer of 27. The answer is 3. Note: $\sqrt[3]{27}$ can also be written as $(27)^{1/3}$.

Exercise Set 15.1

Involve the students in finding the answers for the following exercises based on the principle of students' differential knowledge bases. By this time, students should have fully mastered perfect squares and square roots. If the students' responses are not automatic and accurate, then some review will be required.

Mentally solve for the unknown quantity.

1. $5x = 35$
2. $9x = 81$
3. $(5n)(3) = 45$
4. $90 = (3p)(5)$

Perform the indicated operations.

1. 4^2 =
2. 5^2 =
3. $(11)^2$ =
4. $(121)^{1/2}$ =
5. $(100)^{1/2}$ =
6. $(10{,}000)^{1/2}$ =
7. $\sqrt{1000000}$ =
8. $(81)^{1/2}$ =
9. $(144)^{1/2}$ =
10. 4^3 =
11. 5^3 =
12. 2^3 =
13. $(27)^{1/3}$ =
14. $(64)^{1/3}$ =
15. $(125)^{1/3}$ =

FORMULAS FOR PERIMETER, AREA, AND VOLUME OF FAMILIAR SHAPES

We will only consider some selected familiar shapes in calculating the perimeter, area, and volume of these shapes. These formulas are listed in table 15.1.

Either list these formulas on the chalkboard or prepare a handout for them. The formulas must be grouped by perimeters, areas, and volumes, and the name of the figure listed against each of the math equations. Try to derive the formulas inductively, at least for a rectangle, square, and rectangular solid.

Before asking your students to do any of the problems in exercise set 15.2, they need to know two other important features about dealing with circles and the area and volume of various shapes. The first feature is that the radius of a circle is one-half the diameter, and the diameter of a circle is twice its radius. The second is about the units of measure, in short units, of perimeter, area, and volume. The unit of the perimeter is the same as that of the base quantity, i.e., length. The units of the area and volume are the *square* and *cube*, respectively, of the unit of the base quantity.

For example, if the basic quantity of length (or the radius) is in cm, then the area will have the unit of cm^2, and the volume of cm^3. Similarly, if the basic quantity of length (or the radius) is in inches (in), then the area will have the unit of in^2 and the volume of in^3.

Table 15.1 Perimeter, Area, and Volumes of Familiar Shapes

Shape	Dimensions	Perimeter	Area	Volume
Rectangle	L = length W = width	P = 2(L + W)	A = LW	
Square	L = length	P = 4L	A = L²	
Rectangular solid	L = length W = width H = height			V = LWH
Cube	L = length			V = L³
Triangle	b = base h = height a = side 2 b = side 3	P = (a + b + c)	$A = \dfrac{bh}{2}$	
Circle	r = radius d = diameter	P = 2πr	A = πr²	
Sphere	r = radius			$V = \dfrac{4\pi r^2}{3}$
Right Circular Cylinder	r = radius h = height		Area of the ends, A = πr²	V = pr²h
Circular cone	r = radius h = height			$V = \dfrac{\pi r^2 h}{3}$

How Big? 195

Equations in Groups

Table 15.2   Equations for Perimeter, Area, and Volume Grouped by Geometrical Shape

Have your students copy this table and place it under their noses when doing problems in this chapter.

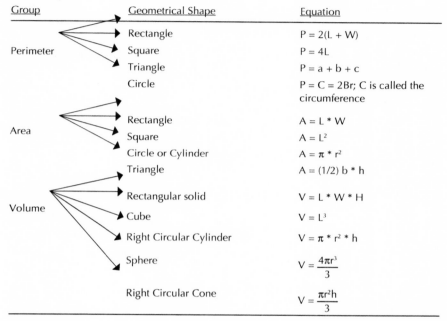

Group	Geometrical Shape	Equation
Perimeter	Rectangle	$P = 2(L + W)$
	Square	$P = 4L$
	Triangle	$P = a + b + c$
	Circle	$P = C = 2Br$; C is called the circumference
Area	Rectangle	$A = L * W$
	Square	$A = L^2$
	Circle or Cylinder	$A = \pi * r^2$
	Triangle	$A = (1/2) b * h$
Volume	Rectangular solid	$V = L * W * H$
	Cube	$V = L^3$
	Right Circular Cylinder	$V = \pi * r^2 * h$
	Sphere	$V = \dfrac{4\pi r^3}{3}$
	Right Circular Cone	$V = \dfrac{\pi r^2 h}{3}$

PERIMETER, AREA, AND VOLUME PROBLEMS WITH SOLUTIONS

Problem 1: The area of a square park is 900 ft^2. How long is one side of the park?

Step 1. Look for a formula for the area of a square. The formula you will find is: $A = L^2$.

Step 2. In the problem, the area A is given as 900 ft^2, and you are required to find the length, L.

Step 3. Set up the equation as $L^2 = 900$. Now remember and recognize that whenever you see an equation containing the square of an unknown quantity as the last step, take the square roots of both sides of the equation.

Step 4. $\sqrt{L^2} = \sqrt{900}$. Remember, the square root of a quantity that is squared is the quantity itself—in this case, L. So, L = 30 ft, because the square root of 900 is 30.

The overall solution will look like:

$$A = L^2$$
$$900 = L^2$$
$$\sqrt{900} = \sqrt{L^2}$$
$$30 \text{ ft} = L$$
or
$$L = 30 \text{ ft}$$

Problem 2: The area of a rectangular park is 1200 ft². The width of the park is 30 ft. What is the length of the park?

Step 1. Look for a formula for area of a rectangle. The formula you will find is: $A = LW$.

Step 2. In the problem, the area, A, is given as 1200 ft², and the width, W, is given to be 30 ft. You are required to find the length, L.

Step 3. Set up the equation as $A = LW = 1200$.

Step 4. Substitute the value of 30 for W, and rewrite the equation as $30L = 1200$.

Step 5. Divide both sides by 30, the coefficient of L.

$$\frac{30L}{30} = \frac{1200}{30}$$

Simplification will give an answer of 40 ft for L.

Problem 3: The perimeter of a square room is 60 ft. How long is one side of the room?

Step 1. Look for a formula for the perimeter of a square. The formula you will find is: $P = 4L$.

Step 2. In the problem, the perimeter, P, is given as 60 ft. You are required to find the length, L, of one side of the room.

Step 3. Set up the equation as $P = 4L = 60$.

Step 4. Divide both sides by 4, the coefficient of L.

$$\frac{4L}{4} = \frac{60}{4}$$

Simplification will give an answer of 15 ft for L.

Problem 4: The perimeter of a rectangular pool is 100 ft. One side of the pool is 30 ft. What is the other dimension of the pool?

Step 1. Look for a formula for the perimeter of a rectangle. The formula you will find is: $P = 2(L + W)$.

Step 2. In the problem, the perimeter, P, is given as 100 ft, and one side's length, L, is given to be 30 ft. You are required to find the other side of the pool, W.

Step 3. Set up the equation as $P = 2(30 + W) = 100$.

Step 4. Divide both sides by 2 first.

$$\frac{2(30 + W)}{2} = \frac{100}{2}$$

Simplification gives $30 + W = 50$.

Step 5. Transpose 30 to the right side of the equal sign, and then change the sign of the term.

$$W = 50 - 30 = 20 \text{ ft}$$

Problem 5: The perimeter of a rectangular public park is 300 ft. The long side of the park (its length) is twice its width. What are the measurements of the two sides of the park?

Step 1. Look for a formula for the perimeter of a rectangle. The formula you will find is: $P = 2(L + W)$.

Step 2. In the problem, the perimeter, P, is given as 300 ft, and the length is twice its width, meaning $L = 2W$. You are required to find the width, W, and from that, the length, L.

Step 3. Set up the equation as $P = 2(L + W) = 2(2W + W) = 300$

Step 4. Divide both sides by 2 first.

$$\frac{2(2W + W)}{2} = \frac{300}{2}$$

This simplifies to $2W + W = 150$.

Step 5. Adding the left hand side of the equation will result in 3W. Therefore, $3W = 150$.

Step 6. Divide both sides by the coefficient of W, which is 3. This division will result in the answer of $\frac{3W}{3} = \frac{150}{3} = 50$ ft for W.

Step 7. The length is twice the width, or $L = 2W$. Since the width W is 50 ft, the length will be $2 * 50 = 100$ ft.

Problem 6: Find the volume in terms of π of a cone with a radius of 4 cm and a height of 10 cm.

Step 1. Look for a formula for the volume of a cone. The formula you will find is:

$$V = \frac{\pi r^2 h}{3}$$

Step 2. In the problem, the radius, r, is given as 4 cm, and the height, h, is 10 cm. You are required to find the volume, V.

Step 3. Substitute the values of radius and height in the equation for volume. You will get

$$V = \frac{\pi 4^2 \cdot 10}{3}$$

Step 4. Square 4, which is 16. Then multiply 16 by 10: 16 × 10 = 160. Many students will forget to square 4 and simply multiply 4 and 10. *Tell your students to be careful.*

Step 5. The final answer for the volume will be $V = \dfrac{160\pi}{3}$ cm³.

Problem 7: Find the volume of a sphere in terms of π with a radius of 5 cm.

Step 1. Look for a formula for the volume of a sphere. The formula you will find is:

$$V = \frac{4\pi r^3}{3}$$

Step 2. In the problem, the radius, r, is given as 5 cm. You are required to find the volume, V.

Step 3. After substituting the values of the radius, you will get

$$V = \frac{4\pi 5^3}{3}$$

Step 4. Take the *cube of 5*, which is 125. Then multiply 125 by 4: 125 × 4 = 500. Many students forget to take the cube of 5; rather, they will simply multiply 5 and 4. Again, *instruct your students to be careful.*

Step 5. The final answer for the volume will be $V = \dfrac{125 \cdot 4\pi}{3} = \dfrac{500\pi}{3}$ cm³.

Exercise Set 15.2

1. Find the radius of a circle whose diameter is 26".
2. Find the diameter of a circle whose radius is 9 in.
3. Find the perimeter of a triangle if the three sides of the triangle are 5", 6", and 9".
4. Find the perimeter and the area of a square whose one side is 9 in.
5. Find the perimeter and the area of a rectangle whose two sides are 9 in and 11 in.
6. Find the circumference of a circle in terms of π whose radius is 9".
7. Find the radius of a circle whose circumference is 40π in.
8. Find the area of a circle in terms of π whose radius is 10 cm.
9. One side of a square measures 15 cm. What are its perimeter in cm and its area in cm²?
10. The area of a square tank is 81 ft². What is one side of the tank?
11. The area of a square park is 400 ft². What is one side of the park?
12. The area of a rectangular park is 1500 ft². The width of the park is 30 ft. What is the length of the park?

13. The perimeter of a square room is 80 ft. What is one side of the room?
14. The perimeter of a rectangular pool is 100 ft. One side of the pool is 30 ft. What is the other side?
15. The perimeter of a rectangular public park is 400 ft. The length side of the park is three times its width. What are the lengths of the two sides of the park?
16. Find the volume of a cone with a radius of 5 cm and height of 20 cm in terms of π.
17. Find the volume of a sphere with a radius of 3 cm in terms of π.
18. Find the volume of a cylinder in cm^3 whose radius is 5 cm if it has a height of 10 cm.
19. Find the area of a triangle in square inches whose base is 10 in and which has a height of 5 in.
20. Find the volume in terms of π of a sphere with a radius of 10 cm.

CHALLENGING PROBLEMS

1. The area of a square park is 1500 ft^2. Is one side of the park less or more than 40 ft?
2. Lisa is tiling a bathroom floor. She needs 12 tiles to cover each square foot of floor space. How many tiles does she need to cover 30 sq ft of floor space?
3. The area of a square playing field is 1,000,000 ft^2. What is one side of the field?
4. The area of a square park is 10000 ft^2. How long is one side of the park?
5. Find the volume of a cylinder whose base diameter is 20 cm and which has a height of 20 cm. You can express the answer in terms of π.
6. Find the volume of a cone with a base diameter of 20 cm and a height of 30 cm. Express the answer in terms of π.
7. Find the volume of a sphere whose diameter is 6 cm. Express the answer in terms of π.

SAMPLE EXIT QUESTIONS

1. Solve for x: $y = mx + b$
2. Solve for y: $y^2 = 81$
3. Solve for z: $z^3 = 125$
4. Solve for r: $V = \frac{4}{3}\pi r^3$
5. Find the volume in terms of π of a sphere whose radius is $\sqrt[3]{3}$.

SAMPLE PRIMING HOMEWORK

1. Simplify: $\dfrac{-5-7}{-2-1}$
2. Identify the symbols in the equation: $y = mx + b$
3. What are the slope and the y-intercept in $y = -3x - 7$
4. Solve for y: $y - 3 = -2(x - 4)$
5. Find the value of the slopes for: $y = -5x - 11$ and $y = -5x + 6$. What can you tell about the slopes of these two lines and the lines themselves?
6. Find the value of the slopes for: $y = 3x - 7$ and $y = -\dfrac{x}{3} + 1$. What can you tell about the slopes of these two lines and the lines themselves?

POINTS TO REMEMBER AND REVIEW

- Watch for the squares and cubes of terms.

    ✓ The square root of a term that is squared is the term itself.
    ✓ The cube root of a term that is cubed is the term itself ($\sqrt[3]{x^3} = x$).

- If you come across a term that is *squared* toward the end of a problem, or if the problem starts with the square, take the *square root* of both sides of the equation to get the final answer.
- If you come across a term that is *cubed* toward the end of a problem, or if the problem starts with the cube, take the *cube root* of both sides of the equation to get the final answer.
- A cross product or cross multiplication is between terms that are diagonally opposite to each other across the equal sign.

16
Slopes and Intercepts on the Algebra Trail

As we know, many students groan when they hear the word "graphs," especially when the slope of a line and intercepts are involved. As teachers, though, we know that students need to learn and practice these in order to remember and apply them.

When your students think of linear graphs, they should be thinking of the coordinate system, the four quadrants, and the coordinates of a point in any of the quadrants. Explain these basic and relevant concepts to your students first, and they must understand and practice them. Only then can we go on to introduce the concepts of slopes and intercepts.

THE COORDINATE SYSTEM

The x- and y-Axes

In this chapter, before you do anything else, your students need to become familiarized with the coordinate system containing the x- and y-axes. The axes pointing north and east are positive, whereas the axes pointing south and west are negative. Such a coordinate system is also known as the *rectangular system of coordinates*. It is divided into four spaces. We label these spaces as the *quadrants*. A complete coordinate system with all these features is shown in graph 16.1.

The Four Quadrants of the Coordinate System

The quadrants are the spaces delineated by the east and north (I quadrant), the north and west (II quadrant), the west and south (III quadrant), and the

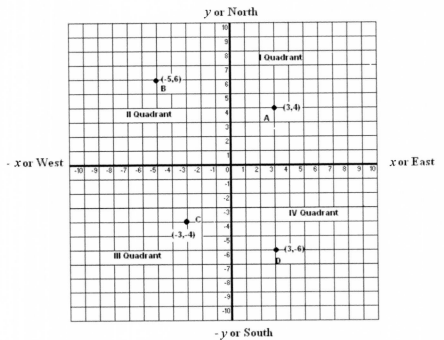

Graph 16.1 Four Quadrants of a Rectangular Coordinate System

south and east (IV quadrant) directions. The *x*- and *y*-values are both positive in the I quadrant. The *x*-value is negative and the *y*-value is positive in the II quadrant. Both values are negative in the III quadrant. The *x*-value is positive and the *y*-value is negative in the IV quadrant. The locations of the four quadrants are shown in graph 16.1.

The Coordinates of a Point

The coordinates of a point on a graph located in relation to its axes are its *x*- and *y*-values. *The x-value is always written first, and the y-value is second.* A comma separates the two values, and the coordinates are enclosed in parentheses.

To illustrate, the coordinate of a certain point, B, (see graph 16.1) can be written as (–5, 6). The *x*-value of that point is –5, and its *y*-value is 6. The negative value of *x* (–5) indicates that the point could be either in the II or the III quadrant. The positive value of *y* (6), however, limits this point to the II quadrant because y-values cannot be positive in the III quadrant.

If the coordinates of a point include a zero (0) as the value of either the *x*- or the *y*-coordinate, then the point is in *none* of the quadrants.

Exercise Set 16.1

In which quadrant, if any, are the points located?

1. (5, 4)
2. (−7, −4)
3. (9, −7)
4. (−6, 6)
5. (0, −3)
6. (4, 6)
7. (7, −4)
8. (4, 4)
9. (0, 0)
10. (−9, −7)
11. (6, −6)
12. (−3, 0)

13. Find coordinates of all the points as marked in graph 16.2.

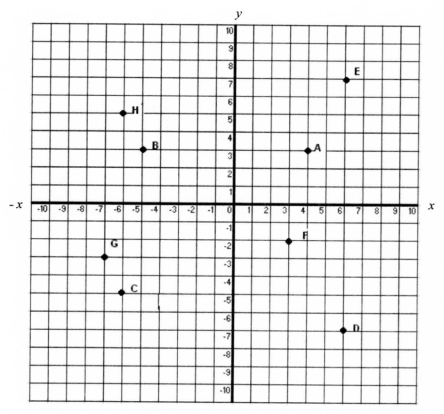

Graph 16.2 Associating Points with Quadrants

On the Graph 16.2A plot the points whose coordinates are:

14. (4, 6)
15. (−3, 5)
16. (−5, −7)
17. (5, −8)
18. (0, 3)
19. (5, 0)

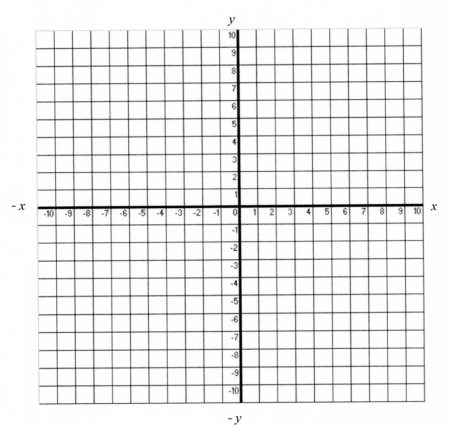

Graph 16.2A Associating Quadrants with Points

SLOPE AND INTERCEPT OF A LINE

Just as a hill has an upward slope (climbing, positive slope) or a downward slope (coming down, negative slope), a line can have a positive or negative slope. The rest of this section deals with the slopes of lines and their intercepts.

The Positive Slope

Most people understand the meaning of the phrase "positive slope," and they can demonstrate it with a rising hand. Ask your students to do the same.

Ask the class to draw the growth chart of a child. One student in the class might draw a line passing through the origin. Another could draw a graph with an intercept on the *x*-axis. Still another student could draw a graph where the height of a child actually decreases with age.

Have patience, because someone in the class will probably draw a graph where the line will intersect the positive *y*-axis. Each time, make an instructional comment that enhances learning. For example, when a student draws a line graph passing through the origin, ask the student, "Were you born with no height? In other words, were you ever zero inches tall when you were born?" Hopefully, the student will laugh, rethink the problem, and then draw the right kind of graph.

Graph 16.3 is such a graph, showing the growth of child in inches with age in years, beginning with birth. The graph shows that the child is born with a certain height, and the height reaches a maximum after a certain age; the child, now an adult, grows no taller. When *both* quantities, height and age, are *increasing*, you have a *positive slope*, and the height and age are said to be *proportional* to each other.

Exercise Set 16.2

Use graph 16.3 to answer the following questions.

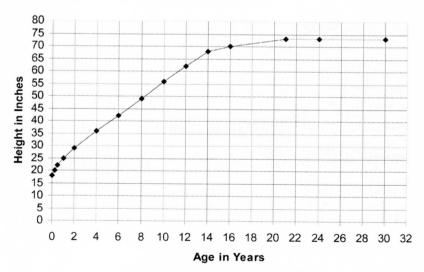

Graph 16.3 Height of a Child vs. Age

1. What is the approximate height of a child at birth?
2. What is the approximate height when the child is four years old?
3. What is the approximate height when the child is sixteen years old?
4. What is the approximate height when this child grows up to twenty-four years old?
5. What is the approximate height when this child reaches thirty years old?

The *y*-Intercept

The *y-intercept* is the *y*-coordinate of a point where the line intersects the *y*-axis and the *x-coordinate is zero*. For example, in graph 16.3, the *y*-intercept is approximately 18 inches, the typical height of a newborn male child. *Students should keep in mind that the x-coordinate of a point must be equated to zero when finding the y-intercept.*

The Negative Slope

As with the case of positive slope, most people understand the meaning of the phrase "negative slope," and they can demonstrate it with one hand going down. Ask your students to do the same.

As mentioned earlier, the negative slope can be associated with walking or skiing down a hill. To illustrate this case, ask your students to

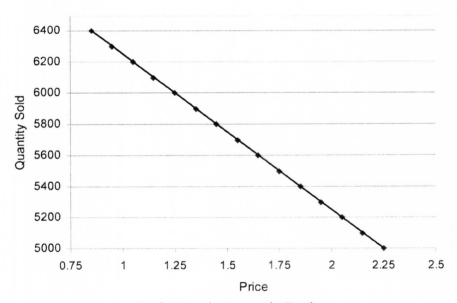

Graph 16.4 Advertisement by Retailers

plot a line graph for the quantity of hamburgers sold when the price is reduced. A hamburger chain is selling two burgers for the price of one as a promotion for attracting new customers. After a few tries, students are going to get it right. They might even surprise you by getting it right the first time.

This situation is depicted in graph 16.4, where the price (in this case, the independent variable, allowing you have complete control over its values) is plotted on the x-axis, and the quantity sold (the dependent variable here, contingent on whether or not customers choose to participate) is plotted on the y-axis.

The graph shows that the quantities of hamburgers sold increase as the price decreases, but no free hamburgers, please. When one quantity, like price, is *dropping*, and the other one, like quantities sold, is *increasing*, we have a *negative* slope.

Exercise Set 16.3

Use graph 16.4 for the problems below.

1. Describe in your own words the independent and dependent variables.
2. What was the approximate price when the quantity of hamburgers sold was 6000?
3. What was the approximate price when the quantity of hamburgers sold was 6300?
4. What was the quantity of hamburgers sold when the price of a hamburger was about $1.75?

The x-Intercept

The x-intercept is the x-coordinate of a point where the line intersects the x-axis and the *y-coordinate is zero*. Again, students need to keep in mind that *the y-coordinate of a point must be equated to zero when finding the x-intercept.*

EQUATION OF A STRAIGHT LINE

Ask your students if anyone can give an equation of a straight line. Brace yourself for a pleasant surprise if some of them seem to know this fact. In any case, ask them to write one on the board and to name the variables involved in the equation. They might surprise you again by naming these variables.

In any case, here it is—the equation of a straight line and the name of the variables in it:

$y = mx + b$,
where
y and x are the y- and x-values,
m = slope of the line, and
b = y-intercept of the line

We will illustrate the use of this equation by providing two examples.

The first example is a graph of force (an independent variable) versus acceleration (a dependent variable) and is linear with a positive slope (see graph 16.5).

To find the slope in graph 16.5, we need two points on a line and their coordinates. The coordinates of point B are (8, 28), whereas the coordinates of point A are (2, 10). *The choice of the points A and B is quite arbitrary.* Here we will label (x_2, y_2) as the coordinates of point B, the second point, and (x_1, y_1) as the coordinates of point A, the first point. In this notation, $x_2 = 8$, $x_1 = 2$, $y_2 = 28$, and $y_1 = 10$.

The *slope* of a line, m, is written as:

$$m = \frac{y_2 - y_1}{x_2 - x_1},$$

and it is *independent of the order* in which the points are chosen. Substituting the x- and y-values in the equation for the slope, we obtain

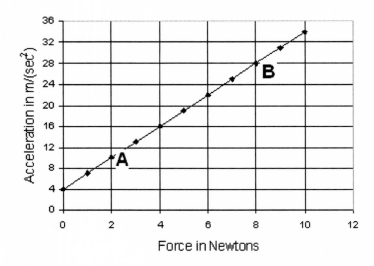

Graph 16.5 *A Straight Line Graph with Positive Slope*

$$m = \frac{28-10}{8-2} = \frac{18}{6} = 3$$

for the slope of the line.

The *y-intercept*, *b*, is equal to 4 because it is the point whose coordinates are (0, 4), that is, where the line intersects the *y*-axis with $x = 0$.

The second example is with retarding force (an independent variable) versus deceleration (a dependent variable) and has a negative slope (see graph 16.6). We need to find the slope and the *y*- or *x*-intercepts.

In graph 16.6, the coordinates of point A are (3, 20), and the coordinates of point B are (9, 8). Using the equation for the slope, as before, we get

$$m = \frac{8-20}{9-3} = \frac{-12}{6} = -2$$

for the slope of the line.

The *x*-intercept is the value of the point on the *x*-axis where $y = 0$; the line intersects the *x*-axis at that point. This point has the *x*-value of 13, meaning that the *x*-intercept is equal to 13; the coordinates of this point are (13, 0).

Exercise Set 16.4

Determine the slope of a line given two pairs of points. *Your students should be able to do these exercises by themselves after you have explained and illustrated the previous section.*

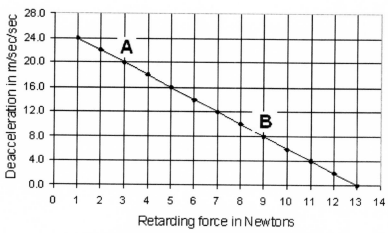

Graph 16.6 A Straight Line Graph with Negative Slope

1. (8, 2) and (2, 6)
2. (2, 4) and (3, 5)
3. (6, 2) and (12, 6)
4. (2, 4) and (3, 6)
5. (4, 4) and (3, 6)
6. (−4, 3) and (6, −7)

Finding the Slope and Intercept—Examples

Place the following examples on the board; then explain them before assigning the exercises that follow. This stuff is tricky at best, so everyone needs to be careful.

Example 16.1. Finding Slope and Intercept

	Example	Explanation
1.	Find the slope and y-intercept of a line given by: $y = 2x + 3$	Compare this equation with the standard equation of a line, $y = mx + b$. A comparison will reveal that m, slope of the line, is 2 and the y-intercept or just the intercept, b, is 3.
2.	Find the slope and y-intercept of a Line given by: $4x + 5y + 20 = 0$	First, reduce the equation to the standard form of a line equation by solving for y; transferring $4x$ and 20 on the right-hand side of the equal sign, e.g., $5y = -4x - 20$. Now divide both sides of the equation with 5, the coefficient of y. This division will result into, $y = -(4/5)x - 4$. Therefore, the slope of this line is $-(4/5)$ and the intercept is -4.
3.	Find the y-intercept of a line: $5y = 10 - 2x$	The y-intercept of a line is its y-value when $x = 0$. Substituting $x = 0$ in the equation, we will get, $5y = 10$. Dividing both sides of the equation by 5 will give $y = 2$, the y-intercept.
4.	Determine the x-intercept of a line: $5y + 4x = 20$	The x-intercept of a line is its x-value when $y = 0$. Substituting $y = 0$ in the equation, we will get, $4x = 20$. Dividing both sides of the equation by 4 will give $x = 5$, the x-intercept.
5.	Determine whether a given point is a solution of the equation: $(1, 7)$, $y = 2x + 5$	Substitute 7 for y and 1 for x and verify the identity, i.e., when both sides have equal values. Substitution will give us: $7 = (2)(1) + 5$ $7 = 2 + 5$ $7 = 7$ The terms on both sides of the equal sign have equal values. Therefore, the point, $(1, 7)$, is a solution.

Exercise Set 16.5

Determine if a given point is a solution of the equation.

1. $(2, 9), y = 2x + 5$
2. $(3, 11), y = 2x + 5$

Find the slope and y-intercept of a line.

3. $y = 4x + 6$
4. $10x + 5y + 30 = 0$
5. $y = -4x - 9$
6. $-6x = 4y + 2$
7. $y = 3x + 5$
8. $4x + 4y + 20 = 0$
9. $y = -4x - 9$
10. $-8x = 4y + 4$

Determine the x-intercept of a line. *Hint:* Let $y = 0$, and then solve the equation for x.

11. $5y + 10x = 30$
12. $5x - 4y = 20$
13. $10y + 4x = 20$
14. $4x - 20y = 20$

Find the *y*-intercept of a line. *Hint:* Let $x = 0$, and then solve the equation for y.

15. $y = 3x$
16. $5y = 10 - 2x$
17. $3y = 18x$
18. $10y = 10 - 2x$

Finding the Equation of a Straight Line

Now, we deal with a special case of finding the equation of a straight line when the slope of that line and the coordinates of one point on the line are given. Let us illustrate this concept with two examples.

Example 16.2. Finding the Equation of a Line

	Example	Explanation
1.	Find the equation of a line when the slope and coordinates of one point are given as: (2, 5), m= 5	Here the coordinates of a point are given to assist you in finding the value of the intercept, b. The general equation of the straight line is, $y = mx + b$. Now substitute 2 for x, 5 for y, and 5 for m, and then solve for b. After substituting these values and rearranging terms, we have, $5 = 5(2) + b$ or $5 = 10 + b$ $5 - 10 = b$ or $b = -5$ To find the equation of a line, all you have to do now is to substitute the value of m and b in the general equation. Since $m = 5$ and $b = -5$, $y = mx + b$ becomes $y = 5x - 5$ for the equation of a straight line.
2.	Find the equation of a line when the slope and coordinates of one point are given: (-3, 0), $m = -2$	As before, substitute -3 for x, 0 for y, and -2 for m in the equation of the straight line. This will result in: $0 = (-2)(-3) + b$ $0 = 6 + b$ $b = -6$ With this value for b, the equation of the straight line will become, $y = -2x - 6$

Exercise Set 16.6

Find the equation of a line when the slope and coordinates of one point are given. *Tell the students to keep the answers in fractions if they are unable to simplify them mentally.*

1. (3, 5), $m = 5$
2. (−4, 0), $m = −2$
3. (4, 3), $m = 3/4$
4. (3, 7), $m = 5$
5. (−3, 0), $m = −2$
6. (2, 4), $m = −2$
7. (3, 5), $m = 6$
8. (−3, 0), $m = − 8$
9. (2, 4), $m = −3$

PARALLEL AND PERPENDICULAR LINES

Ask your students to draw perpendicular (vertical) and horizontal lines on the board, or ask them to illustrate these two types of lines by hand. Next, ask them about the slopes of these lines. Some of the students may come up with the right answer. Remind them to recall the property of zero (0) when used in the numerator or denominator of an expression. *The slope of a line parallel to the y-axis is infinity, whereas the slope of a line parallel to the x-axis is zero.* The explanation for these statements is as follows, and once the explanation for these answers is provided, students should commit these facts to memory.

When we deal with any two lines, each of them will have a slope. Let us denote the slope of the first line by m_1 and the slope of the second by m_2. Based on this information, we can draw two conclusions about these two lines.

If the two slopes are equal, (i.e., $m_1 = m_2$), the two lines are parallel. If $m_1 * m_2 = -1$, then the two lines are perpendicular. For example, if one slope, m_1, is 3, the other slope, m_2, has to be -1/3 for the lines to be perpendicular.

Long extended explanations about these conclusions are unnecessary here. Nevertheless, if some of your students want to verify these conclusions, they can do so by solving various types of problems. You should encourage them to find or make up their own examples that will lead to these conclusions.

Equations for Parallel and Perpendicular Lines—Examples

	Statement	Explanation
1.	The slope of a line parallel to the y-axis (the vertical line) is undefined. 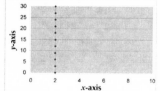	The slope of a line is defined as $m = \dfrac{y_2 - y_1}{x_2 - x_1}$. For a line parallel to the y-axis, the two x-values are the same. Therefore, their x-difference $(x_2 - x_1)$ is zero. Dividing a quantity by zero will lead to answer of <u>undefined</u> for the slope.
2.	The slope of a line parallel to the x-axis (the horizontal line) is zero. 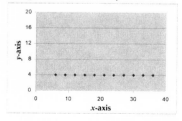	The slope of a line is defined as $m = \dfrac{y_2 - y_1}{x_2 - x_1}$. For a line parallel to the x-axis, the two y-values are the same. Therefore, their difference $(y_2 - y_1)$ is zero. Dividing zero by any quantity will give us an answer of <u>zero</u> for the slope.

Example 16.3 Explanation of Perpendicular and Parallel Lines

Let us warn you ahead of time that these examples are difficult and sometimes quite tricky. The only way to make it easy on students is to ask them to solve as many problems as they can and to ask you for assistance if they become baffled and frustrated.

Exercise Set 16.7

1. Find an equation of a line that contains the point (0, 3) and is parallel to $y - 6x = 6$.
2. Find an equation of a line that contains the point (0, 4) and is perpendicular to $3y - x = 0$.
3. Find the value of k such that the lines given by $6y = kx - 12$ and $4x + 12y = 24$ are perpendicular.
4. Find the value of k such that the lines given by $8y = kx - 8$ and $4x + 8y = 24$ are parallel.

MENTAL CHAPTER REVIEW

First, have your students correctly label the quadrants of the graph before they proceed. Then, ask them to find coordinates of the points as marked in the x-y grid, graph 16.7.

In which quadrant, if any, are the following points located?

1. (6 , 4)
2. (−7, −5)
3. (11, −7)
4. (−6, 7)
5. (0, −4)
6. (5, 6)
7. (6, −4)
8. (6, 4)
9. (0, 0)
10. (−10, −7)
11. (7, −6)
12. (−5, 0)

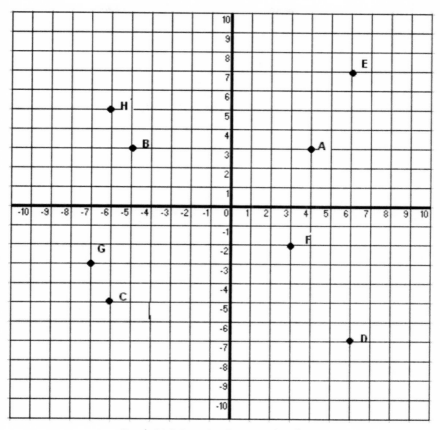

Graph 16.7 Locating Points in Quadrants

Identify the quadrant and mark the points on the grid below graph 16.8, whose coordinates are:

1. (6, 6)
2. (−5, 5)
3. (−6, −7)
4. (6, −8)
5. (0, 2)
6. (4, 0)

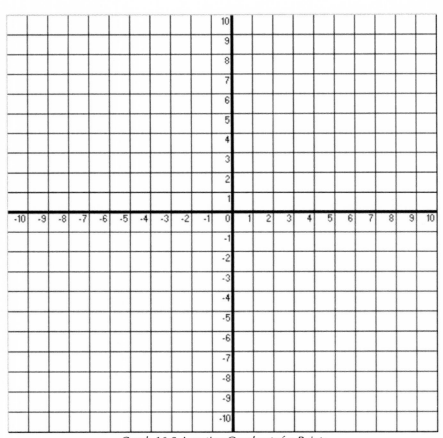

Graph 16.8 Locating Quadrants for Points

Perform the indicated operations.

1. $\dfrac{312}{345-345} =$

2. $\dfrac{10-(-5-4)}{21+(-2)} =$

3. $12 - (-5 - 2) + (5 - 6) =$

4. $\dfrac{20-(7+17)}{21+(-23)} =$

5. $\dfrac{11-(18+15)}{-12+23} =$

SAMPLE EXIT QUESTIONS

1. Find the slope of a line passing through the points $(-1, -3)$ and $(3, 5)$.
2. Find the slope and the y-intercept of the line $3y = -21x + 8$.
3. Find the x-intercept of the line $y = x$.
4. Plot the line $x = 5$.

SAMPLE PRIMING HOMEWORK

Simplify:

1. $\dfrac{1}{4} + \dfrac{3}{4}$

2. $2\dfrac{3}{4} + \dfrac{3}{4}$

3. $5\dfrac{1}{2} - 3\dfrac{1}{2}$

4. $11\dfrac{1}{2} - 5\dfrac{1}{2}$

5. $\dfrac{1}{2} \div 2$

6. $3\dfrac{3}{5} * \dfrac{5}{9}$

POINTS TO REMEMBER AND REVIEW

- Write equations in the standard form, $y = mx + b$, meaning the coefficient of y has to be *one* and y has to be *by itself on one side* of the equation before finding the slope or the y-intercept.
- If the constant b is missing, the y-intercept is zero.
- When looking for the y-intercept, make $x = 0$, and solve for y.
- When looking for the x-intercept, make $y = 0$, and solve for x.
- When looking for an equation, remember the general equation: $y - y_1 = m(x - x_1)$. In this equation, you only substitute values for y_1 and x_1; y and x do not change.

17
Mixed Numerals

For student success in completing the tasks of this chapter, your students should have developed complete mental mastery of simple fractions and their arithmetic operations. Just in case they haven't quite done so, we start the chapter by providing a review of these basics. If your students are well prepared, you can go over the review lightly and really begin the serious work of the chapter with the subsequent section, "Improper Fractions as Mixed Numerals."

REVIEW OF STUDENTS' KNOWLEDGE OF SIMPLE FRACTIONS

To begin this review, we start with a term written as 3/11, a fractional number; 3 is the *numerator* of the fraction, and 11, the *denominator*. Fractions are of two types: a *proper fraction* is one where the numerator is *less* than its denominator, like 3/11; an *improper fraction* is the other type where the numerator is *greater than or equal to* the denominator, as in 11/3.

This chapter can seem to be deceptively simple. After all, the concepts only involve simple addition and subtraction. But the problem-solving process involves many steps, and often students have difficulties with the problems because the process requires them to stay focused on carrying out all of the steps. This chapter is about practice, more practice, and then some more practice.

But there is light at the end of the tunnel. With sufficient practice comes the mastery of complex fractions. When your students reach this stage of mastery, they should be feeling very confident in their ability to stay focused and solve complex fraction problems.

As we stated in an earlier chapter, fractions are all around us. For example, 50 cents is 1/2 of a dollar; 25 cents is 1/4 of a dollar; a dime is 1/10 of a dollar. All these are examples of fractions. When you go to a grocery store, you deal with half (1/2) gallons and quarts (1/4 of a gallon). Recognition of a fraction even before doing an arithmetic operation is important when working with fractions. For example, 16/4 is simply 4; 12/24 is 1/2.

The following operations commonly used with fractions need to be practiced and eventually committed to memory.

Exercise Set 17.1

Continue to quiz your students on a daily basis until all of them have the answers on their fingertips; allow no exceptions. Feel free to use variations of these fractions when helping students master them.

1. $\dfrac{1}{2} + \dfrac{1}{2} = 1$

2. $\dfrac{1}{4} + \dfrac{1}{4} = \dfrac{1}{2}$

3. $\dfrac{1}{2} + \dfrac{1}{4} = \dfrac{2}{4} + \dfrac{1}{4} = \dfrac{3}{4}$

4. $\dfrac{1}{4} + \dfrac{1}{4} + \dfrac{1}{4} + \dfrac{1}{4} = 1$

5. $\dfrac{1}{8} + \dfrac{1}{8} + \dfrac{1}{8} + \dfrac{1}{8} = \dfrac{4}{8} = \dfrac{1}{2}$

6. $\dfrac{1}{2} \div 2 = \dfrac{1}{2} \div \dfrac{2}{1} = \dfrac{1}{2} \times \dfrac{1}{2} = \dfrac{1}{4}$

7. $\dfrac{1}{2} \times 2 = \dfrac{1}{2} \times \dfrac{2}{1} = 1$

8. $\dfrac{1}{4} + \dfrac{3}{4} = 1$

9. $\dfrac{1}{5} + \dfrac{4}{5} = \dfrac{5}{5} = 1$

10. $\dfrac{1}{10} + \dfrac{9}{10} = \dfrac{10}{10} = 1$

IMPROPER FRACTIONS AS MIXED NUMERALS

Again, your students probably come across these types of fractions on a daily basis. For example, five quarters can be written as $\frac{5}{4}$, which is an *improper fraction*. That is, *the numerator in this example is larger than the denominator*. However, students will easily recognize that five quarters is a dollar and a quarter—as a fraction: $1\frac{1}{4}$. This last version is called a *mixed numeral*, made up of an integer and a fraction. As a matter of fact, the fractions $\frac{5}{4}$ and $1\frac{1}{4}$ are *equivalent*; they are just written in different ways.

Your students might wonder how you arrived at all these variations. Tell the students there are certain rules that they can follow that will allow them to write these fractions interchangeably. But before we introduce the rules, let your students practice with some simple ones.

Exercise Set 17.2

Introduce your students to the following fractions in an inductive way. Ask them to write the answers to the following problems: first, as a mixed numeral, and then as an improper fraction. If they cannot do it, lead them to the answers. They need to see these operations in the same way the experts see them.

1. $\frac{1}{2} + 1 = 1\frac{1}{2} = \frac{3}{2}$

2. $\frac{1}{2} + \frac{1}{2} + 1 = 1 + 1 = 2$

3. $\frac{1}{4} + \frac{1}{4} + 1 = \frac{1}{2} + 1 = \frac{3}{2}$

4. $\frac{1}{2} + 3 = 3\frac{1}{2} = \frac{7}{2}$

5. $\frac{1}{4} + 5 = 5\frac{1}{4} = \frac{21}{4}$

6. $\frac{3}{5} + 4 = 4\frac{3}{5} = \frac{23}{5}$

Now, we come to the rules we mentioned earlier.

RULES FOR CHANGING AN IMPROPER FRACTION TO A MIXED NUMERAL AND VICE VERSA

Example 17.1. Improper Fractions and Mixed Numerals

	Examples	Explanation
1.	Change an improper fraction, $\frac{17}{5}$, to a mixed numeral	(1) Divide 5 into 17. The process gives 3 with a remainder of 2. (2) The answer is then written as, $3\frac{2}{5}$.
2.	Change a mixed numeral, $6\frac{2}{5}$, to an improper fraction.	(1) Multiply 5 and 6 to give you 30. (2) Add 2 to this product to give you 32. (3) <u>The denominator stays the same</u>. The final answer is $\frac{32}{5}$.
3.	Change an improper fraction, $\frac{26}{8}$, to a mixed numeral and simplify.	(1) Divide 8 into 26. (2) This process gives 3 with a remainder of 2. The answer is written as, $3\frac{2}{8}$. (3) Then simplify the fractional part to give you, $\frac{1}{4}$. The final answer, therefore, is $3\frac{1}{4}$.
4.	Change a mixed numeral, $6\frac{2}{8}$, to a fraction (improper fraction) and simplify	(1) Multiply 8 and 6 to give you 48. (2) Add 2 to this product to give you 50. The answer is, therefore $\frac{50}{8}$. (3) Then, simplification of this fraction will give you the final answer of $\frac{25}{4}$.

Exercise Set 17.3

Change to improper fraction (fractional notation), and simplify if possible.

1. $1\frac{1}{2} =$
2. $1\frac{1}{4} =$
3. $2\frac{3}{4} =$
4. $3\frac{1}{4} =$
5. $4\frac{4}{6} =$
6. $5\frac{4}{9} =$
7. $6\frac{3}{7} =$
8. $9\frac{2}{7} =$
9. $9\frac{1}{6} =$
10. $9\frac{6}{8} =$
11. $9\frac{5}{6} =$
12. $8\frac{5}{8} =$
13. $9\frac{4}{9} =$
14. $8\frac{4}{7} =$
15. $6\frac{2}{8} =$
16. $5\frac{6}{7} =$
17. $10\frac{8}{11} =$
18. $11\frac{9}{11} =$
19. $12\frac{11}{12} =$
20. $7\frac{13}{13} =$
21. $15\frac{5}{15} =$

Simplify if possible, and change to mixed numerals.

1. $\frac{8}{5} =$
2. $\frac{9}{6} =$
3. $\frac{18}{4} =$
4. $\frac{22}{6} =$
5. $\frac{13}{4} =$
6. $\frac{18}{7} =$
7. $\frac{20}{9} =$
8. $\frac{25}{11} =$
9. $\frac{27}{12} =$
10. $\frac{29}{13} =$
11. $\frac{31}{14} =$
12. $\frac{48}{15} =$
13. $\frac{49}{15} =$
14. $\frac{50}{20} =$
15. $\frac{180}{120} =$
16. $\frac{5600}{1100} =$
17. $\frac{23}{5} =$
18. $\frac{450}{150} =$
19. $\frac{90}{40} =$
20. $\frac{67}{8} =$
21. $\frac{58}{8} =$

BASIC MATH OPERATIONS WITH MIXED NUMERALS

In general, students use at least two methods for adding and subtracting mixed numerals. Here we will use one method where your students are less likely to make mistakes. Furthermore, this method is likely to provide the additional practice needed to master these mixed numerals and their math operations. After the practice here, we will be dealing with the multiplication and division of mixed numerals.

Addition and Subtraction of Two Mixed Numerals

Solving problems of this nature is easier and more accurate if your students follow some basic rules. These rules are:

- Change the mixed numerals to improper fractions.
- Simplify, if possible. Find the LCD (least common denominator) or LCM (least common multiple) of the terms, if needed.
- Apply the principles of adding simple fractions with equal and *not* equal denominators.
- Change the answer back to a mixed numeral.
- Simplify the fraction part if you can.

Let us illustrate these rules with the help of examples.

Example 17.2. Adding and Subtracting Mixed Numerals

#	Example: Perform the indicated operations, and write the answer as a mixed numeral	Solution and Explanation: Change the mixed numeral(s) to improper fractions (fractional notation) first, and then follow the other rules.
1.	$2\frac{5}{8} + 3\frac{7}{8}$	$\frac{21}{8} + \frac{31}{8} = \frac{21 + 31}{8} = \frac{52}{8} = 6\frac{4}{8} = 6\frac{1}{2}$
2.	$2\frac{2}{3} + 3\frac{3}{4}$	$\frac{8}{3} + \frac{15}{4} = \frac{32 + 45}{12} = \frac{77}{12} = 6\frac{5}{12}$
3.	$6\frac{5}{8} - 3\frac{7}{8}$	$\frac{53}{8} - \frac{31}{8} = \frac{53 - 31}{8} = \frac{22}{8} = 2\frac{6}{8} = 2\frac{3}{4}$
4.	$7\frac{2}{3} - 3\frac{3}{4}$	$\frac{23}{3} - \frac{15}{4} = \frac{92 - 45}{12} = \frac{47}{12} = 3\frac{11}{12}$
5.	$11 - 3\frac{3}{4}$	$\frac{11}{1} - \frac{15}{4} = \frac{44 - 15}{4} = \frac{29}{4} = 7\frac{1}{4}$

Exercise Set 17.4

Perform the indicated operations, and write the answer as a mixed numeral.
 If the students have any difficulty with performing the task mentally, write the problems on the chalkboard or a flip chart.

1. $7\frac{1}{4} + 3\frac{1}{4} =$
2. $7\frac{1}{4} + 5\frac{2}{4} =$
3. $8\frac{1}{4} + 5\frac{3}{4} =$
4. $5\frac{1}{2} + 3\frac{1}{4} =$
5. $4\frac{1}{8} + 9\frac{3}{8} =$
6. $2\frac{3}{16} + 6\frac{5}{16} =$
7. $8 + 3\frac{1}{4} =$
8. $10 + 7\frac{3}{4} =$
9. $6 + 5\frac{11}{16} =$
10. $7\frac{2}{3} + 11 =$
11. $7\frac{1}{3} - 3\frac{1}{3} =$
12. $9\frac{1}{2} - 4\frac{1}{4} =$
13. $7 - 3\frac{1}{2} =$
14. $11 - 3\frac{3}{4} =$
15. $8 - 5\frac{3}{8} =$
16. $9\frac{2}{3} - 3\frac{4}{6} =$
17. $11\frac{1}{2} - 4\frac{4}{8} =$
18. $14\frac{1}{2} - 11\frac{1}{4} =$
19. $13\frac{2}{3} - 7 =$
20. $15\frac{8}{24} - 9 =$
21. $9\frac{3}{4} - 9\frac{1}{4} =$

Add, and then write the answer as a mixed numeral if appropriate.

22. $2\frac{7}{8} + 3\frac{5}{8} =$
23. $5\frac{5}{6} + 2\frac{5}{6} =$
24. $3\frac{2}{5} + 2\frac{7}{10} =$
25. $4\frac{1}{2} + 2\frac{7}{10} =$
26. $2\frac{3}{4} + 3\frac{5}{6} =$
27. $\frac{3}{4} + 2\frac{5}{6} =$
28. $1\frac{1}{4} + 2\frac{2}{3} =$
29. $3\frac{5}{8} + 4\frac{3}{4} =$
30. $4\frac{1}{4} + 2\frac{7}{8} =$

Subtract, and then write the answer as a mixed numeral if appropriate.

31. $5\frac{1}{5} - 2\frac{3}{5} =$
32. $6\frac{1}{8} - 3\frac{3}{8} =$
33. $5\frac{3}{5} - 2\frac{1}{2} =$
34. $5\frac{2}{3} - 3\frac{1}{2} =$
35. $5\frac{1}{3} - 2\frac{5}{6} =$
36. $5\frac{3}{16} - 2\frac{3}{4} =$
37. $5 - 2\frac{3}{4} =$
38. $8 - 2\frac{3}{8} =$
39. $5\frac{2}{5} - 2 =$

Multiplication and Division of Two Mixed Numerals

In the case of multiplication and division, tell your students to follow the basic rules to solve these kinds of problems. These rules are:

- Change the mixed numerals to improper fractions.
- Place a one (1) in the denominator of a whole number.
- For *multiplication*, simplify and cancel terms.
- For *division*, change the division sign (÷) to the multiplication sign, and then flip-flop (taking the reciprocal—changing numerator to denominator and denominator to numerator) the term that follows the division sign.
- Simplify and cancel terms if you can, but be careful.
- Multiply the remaining numerators.
- Multiply the remaining denominators.
- Simplify if you can.
- Change the answer back to a mixed numeral.

Let us again illustrate these rules with the help of examples.

Example 17.3. Multiplying and Dividing Mixed Numerals

	Example: Perform the indicated operations, and write the answer as a mixed numeral	Solution and Explanation: Change the mixed numeral(s) to improper fractions (fractional notation) first and then follow the other rules.
1.	$3\frac{6}{8} * 1\frac{9}{15}$	$\frac{30}{8} * \frac{24}{15} = \frac{2*3}{1*1} = \frac{6}{1} = 6$. This is because 15 divides into 30 by 2, and 8 divides into 24 by 3.
2.	$2\frac{3}{4} * 2\frac{1}{3}$	$\frac{11}{4} * \frac{7}{3} = \frac{77}{12} = 6\frac{5}{12}$
3.	$4\frac{2}{3} \div 1\frac{3}{4}$	$\frac{14}{3} \div \frac{7}{4} = \frac{14}{3} * \frac{4}{7} = \frac{2}{3} * \frac{4}{1} = \frac{8}{3} = 2\frac{2}{3}$. Here, 7 divides into 14 by 2.

Exercise Set 17.5

Multiply and write the answer as a mixed numeral if appropriate. (Remember that whole numbers have a denominator of 1.)

1. $9 * 2\frac{3}{6} =$
2. $5 * 2\frac{3}{4} =$
3. $4\frac{2}{3} * \frac{1}{4} =$
4. $4\frac{2}{7} * 2\frac{1}{3} =$
5. $2\frac{1}{5} * 2\frac{3}{4} =$
6. $3\frac{1}{5} * 1\frac{7}{8} =$
7. $2\frac{4}{10} * 6\frac{4}{6} =$
8. $4\frac{10}{20} * 5\frac{5}{5} =$
9. $1\frac{6}{10} * \frac{9}{10} =$

Divide and write the answer as a mixed numeral if appropriate. (Remember that whole numbers have a denominator of 1.)

10. $32 \div 3\frac{1}{5} =$
11. $27 \div 2\frac{1}{4} =$
12. $9\frac{4}{5} \div 7 =$
13. $3\frac{3}{8} \div 9 =$
14. $4\frac{1}{4} \div 3\frac{2}{5} =$
15. $5\frac{3}{5} \div 3\frac{3}{6} =$
16. $1\frac{7}{8} \div 1\frac{2}{3} =$
17. $4\frac{2}{8} \div 2\frac{5}{6} =$
18. $5\frac{1}{10} \div 3\frac{2}{5} =$

MENTAL CHAPTER REVIEW

Encourage your students to perform the following operations mentally if they are comfortable with the task.

1. $\frac{1}{4}+\frac{1}{2}=$
2. $\frac{1}{4}+\frac{1}{4}=$
3. $\frac{1}{4}+\frac{3}{4}=$
4. $\frac{1}{4}+1=$
5. $1+\frac{1}{2}=$
6. $\frac{1}{4}+\frac{1}{4}+\frac{1}{4}=$
7. $\frac{1}{4}+\frac{1}{4}+\frac{1}{4}+\frac{1}{4}=$
8. $\frac{1}{4}+\frac{1}{2}+\frac{1}{4}=$
9. $\frac{3}{4}+\frac{1}{4}=$
10. $\frac{1}{2}+\frac{1}{2}+\frac{1}{2}+\frac{1}{2}=$
11. $\frac{1}{2}+\frac{1}{2}+2=$
12. $\frac{1}{2}+\frac{1}{2}+4=$
13. $\frac{1}{4}+\frac{1}{4}+\frac{2}{4}=$
14. $\frac{1}{4}+\frac{1}{4}+\frac{3}{6}=$
15. $\frac{1}{4}+\frac{1}{4}+\frac{8}{16}=$
16. $\frac{1}{4}+\frac{3}{4}+1=$

17. $\frac{1}{4}+\frac{3}{4}+2=$
18. $\frac{1}{8}+\frac{1}{8}=$
19. $\frac{1}{8}+\frac{1}{8}+\frac{1}{8}=$
20. $\frac{1}{8}+\frac{3}{8}+\frac{1}{2}=$
21. $\frac{1}{4}+\frac{1}{2}+\frac{5}{4}=$
22. $\frac{1}{2}*\frac{1}{2}=$
23. $\frac{1}{2}*\frac{1}{4}=$
24. $\frac{1}{2}*\frac{3}{4}=$
25. $\frac{1}{4}*\frac{16}{4}=$
26. $\frac{4}{2}*\frac{8}{16}=$
27. $\frac{1}{4}*\frac{3}{2}*\frac{8}{9}=$
28. $\frac{5}{4}*\frac{16}{4}*\frac{5}{25}=$
29. $\frac{64}{16}*\frac{7}{4}*\frac{6}{21}=$
30. $\frac{9}{8}*\frac{4}{27}*\frac{12}{2}=$
31. $\frac{3}{2}\div\frac{3}{4}=$
32. $\frac{5}{6}\div\frac{25}{30}=$

33. $\frac{1}{2}\div 2=$
34. $\frac{1}{4}\div 4=$
35. $\frac{1}{2}\div\frac{1}{2}=$
36. $\frac{1}{8}\div\frac{8}{64}=$
37. $\frac{1}{2}\div\frac{4}{8}=$
38. $\frac{1}{4}\div\frac{5}{20}=$
39. $\frac{2}{3}\div\frac{12}{18}=$
40. $\frac{3}{2}-\frac{1}{2}=$
41. $1-\frac{1}{2}=$
42. $1-\frac{1}{4}=$
43. $1-\frac{1}{8}=$
44. $2-\frac{1}{2}=$
45. $3-\frac{1}{4}=$
46. $5-\frac{3}{4}=$
47. $\frac{5}{2}-\frac{3}{2}=$
48. $\frac{9}{16}-\frac{5}{16}=$

MENTAL WORD PROBLEMS

49. How much will it cost you for eight candy bars if each bar costs a quarter?
50. You have three wooden blocks. The width of each block is 1/4 of an inch. If you put them together, what will be the total width?
51. Your parents gave you $20.00 for your weekly allowance. You went to a mall with your friends on a mini spending spree. You bought a special drink for $3.25, chocolate candy for $1.25, a paperback book for $3.50, and French fries for $1.00. How much do you still have at the end?
52. You were supposed to attend a birthday party for a friend. You arrived very late for it. By the time you got there, the other guests had consumed 3/4 of the cake. You and your friend ate the rest of the cake in equal amounts. What fraction of the total cake did you eat?
53. You and your two other friends are doing an experiment in measurement. You are assigned to find the volume (length × width × height) of a box. You measure the length to be 2 inches; the two other measurements of width and height by your friends were 1 inch and 1/2 inch. What is the volume of the box in cubic inches?

SAMPLE EXIT QUESTIONS

Simplify:

1. $5\frac{3}{4} + 3\frac{1}{4} =$

2. $5\frac{2}{3} - 2\frac{3}{4} =$

3. $2\frac{3}{8} * 1\frac{5}{19} =$

4. $2\frac{7}{8} * 1\frac{7}{16} =$

SAMPLE PRIMING HOMEWORK

Simplify:

1. $(x^0)(y-z)^0(5p)^0 =$
2. $\dfrac{100 * y^{-8}}{20y^{-11}} =$
3. $(x+4)(x+4) =$
4. $(x-11)(x+11) =$
5. $\left[\dfrac{123x^{11}}{123x^{11}}\right] - (p-100y)^0 + 2*(x^5)(x^{-5}) =$

POINTS TO REMEMBER AND REVIEW

- In multiplying and dividing mixed numerals, first, change them to improper fractions.
- When dividing, change the division sign ($\div$) to a multiplication sign, and then take the reciprocal of the term that follows the division sign.
- Take extra precaution when subtracting two mixed numerals:

  ✓ Subtract the whole numbers first.
  ✓ Second, deal with the fractions using LCM.
  ✓ If the answer happens to be a negative fraction in the previous step, borrowing may be necessary.
  ✓ The final answer is usually positive.

18

Polynomials and Their Basic Operations

In this chapter, we take up polynomials and their basic operations. In algebra, an expression comprised of a single term (e.g., $3x$ or 6) is called a *monomial*. An expression containing multiple terms (e.g., $4x + 5$) is called a *polynomial*.

However, before we attempt to introduce the formal definitions of a monomial and a polynomial, we want to be sure students are fully conversant with exponential notation. If you find your students somewhat familiar with the concepts of exponential notation, quiz them orally on the product and quotient rules involving terms with the same base.

We suggest always checking the students' knowledge bases prior to beginning a new set of concepts or topics to ascertain what they have seen and perhaps even learned in their previous classes. In other words, check their knowledge levels and tailor the review of the knowledge required to succeed with polynomials and their operations. If only a few students possess the knowledge, then a "fishing expedition" of their knowledge bases will be needed.

EXPONENTIAL NOTATION

Here is a review of exponential notation for you to use selectively in preparing your students for this chapter's topics. When we write the number 16 as 2^4, we are writing it in an *exponential form*. In this expression, 2 is the *base* and 4 is the *exponent*.

Now we are ready to provide formal definitions of a monomial and other polynomials. In general, a monomial is an expression of the type

ax^n, where a is a real number and n is a *nonnegative* integer. Some examples of monomials are $5x$, $3x^3$, -7, $121r^4$, and 0. An expression containing the sums or differences of *two monomials* is called a *binomial*. Some examples are $2x - 5$, $5x^3 + 2x$, and $-7b^4 + 9$. Beware of such expressions as $\frac{x+5}{x-3}$; $5x^3 - 4x^2 + \frac{1}{x}$; or $\frac{1}{y^4 - 6}$. They are *not* polynomials because they are not of the form ax^n.

BASIC REVIEW OF EXPONENTS AND INTEGERS AS EXPONENTS

Exponents of 0 and 1

Any term raised to the power of 0 (i.e., having an exponent of 0) is equal to 1. Examples:

$$5^0 = 1$$
$$(246)^0 = 1$$
$$y^0 = 1$$
$$(z^0)(16)^0(p^0) = 1$$

Any term raised to the power of 1 (i.e., having an exponent of 1) is the term itself. Examples:

$$(13)^1 = 13$$
$$(x)^1 = x$$
$$(10)^1 = 10$$
$$(11)^1(x)^1(y)^1 = 11xy$$

Multiplying and Dividing Powers with Like Bases

When multiplying terms in exponential notation with the same base, add the powers (exponents) and keep the base as is. (Remember, the base cannot be zero.) Examples:

$$(x^m)(x^n) = x^{m+n}$$
$$(x^5)(x^7) = x^{12}$$
$$(z^7)(z^{11}) = z^{18}$$
$$(p^{11})(p^{-10}) = p^1 = p$$
$$(q^8)(q^{-9}) = q^{-1}$$
$$(y^{12})(y^{-12}) = y^0 = 1$$

Polynomials and Their Basic Operations

When dividing terms in exponential notation having the same base, subtract the power (exponent) of the denominator from the power (exponent) of the numerator and keep the base as is. (Remember, the base cannot be zero.) Examples:

$$\frac{x^m}{x^n} = x^{m-n}$$

$$\frac{x^{11}}{x^7} = x^{11-7} = x^4$$

$$\frac{x^{11}}{x^{13}} = x^{11-13} = x^{-2}$$

$$\frac{x^{-11}}{x^7} = x^{-11-7} = x^{-18}$$

$$\frac{x^{-11}}{x^{-7}} = x^{-11+7} = x^{-4}$$

$$\frac{x^{12}}{x^{12}} = x^{12-12} = x^0 = 1$$

Negative Exponent

For any real number b that is nonzero and any integer n, $b^{-n} = \frac{1}{b^n}$. (Remember, the base cannot be zero.) Examples:

$$b^{-1} = \frac{1}{b^1} = \frac{1}{b}$$

$$b^{-6} = \frac{1}{b^6}$$

$$b^{-3} = \frac{1}{b^3}$$

$$p^{-9} = \frac{1}{p^9}$$

Raising Powers to Powers: The Power Rule

In raising a power to a power, simply multiply the exponents. For any real number b and any integers m and n: $(b^m)^n = b^{mn}$. Examples:

$$(3^5)^4 = 3^{20}$$
$$(2^5)^{-4} = 2^{-20}$$
$$(5^8)^{1/2} = 5^4$$
$$(9^2)^{1/2} = 9^1 = 9$$

Raising a Product to a Power

To raise a product to the nth power, raise each factor in the product to the nth power. For any real numbers a and b and any integer n, $(ab)^n = a^n b^n$.
Examples:

$$(3x^3)^2 = 3^2 x^6 = 9x^6$$
$$(3a^2 b^3 c^5)^3 = 3^3 a^6 b^9 c^{15} = 27 a^6 b^9 c^{15}$$
$$[(-x)^{13}]^2 = (-x)^{26} = (-1)^{26} x^{26} = x^{26}$$
$$(-3a^3 b^4 c^{-6})^3 = (-3)^3 a^9 b^{12} c^{-18} = -27 a^9 b^{12} c^{-18} = \frac{-27 a^9 b^{12}}{c^{18}}$$

Scientific Notation

Examples:

$$79{,}000 = 7.9 \times 10^4$$
$$0.000035 = 3.5 \times 10^{-5}$$
$$(9 \times 10^5)(4 \times 10^8) = 36 \times 10^{13} = 3.6 \times 10^1 \times 10^{13} = 3.6 \times 10^{14}$$
$$(11 \times 10^5)(11 \times 10^{-9}) = 121 \times 10^{-4} = 1.21 \times 10^2 \times 10^{-4} = 1.21 \times 10^{-2}$$
$$7 \times 10^8 + 9 \times 10^8 = 16 \times 10^8 = 1.6 \times 10^1 \times 10^8 = 1.6 \times 10^9$$
$$13 \times 10^7 - 7 \times 10^7 = 6 \times 10^7$$

Summary of Definitions and Rules for Exponents: $a, b \neq 0$

Exponent of 1:	$b^1 = b$
Exponent of 0:	$b^0 = 1$
Negative Exponents:	$a^{-n} = \dfrac{1}{a^n}$
Product Rule:	$(b^m)(b^n) = b^{m+n}$
Quotient Rule:	$\dfrac{a^m}{a^n} = a^{m-n}$
Power Rule:	$(a^m)^n = a^{mn}$
Raising a product to a power:	$(ab)^n = a^n b^n$
Raising a quotient to a power:	$\left(\dfrac{a}{b}\right)^n = \dfrac{a^n}{b^n}$

Exercise Set 18.1

Evaluate:

1. $(7p)^2 =$
2. $(134.56)^0 =$
3. $(xy)^0(89)^0(11.23)^0 =$
4. m^3 when $m = 3$
5. $y^2 - 7$ when $y = -10$
6. $-b^2$ when $b = -5$
7. $(-b)^2$ when $b = -5$
8. $(-b)^3$ when $b = -2$

Express using positive exponents:

9. $3^{-2} =$
10. $\dfrac{1}{n^{-5}} =$

Express using negative exponents:

11. $4^5 =$
12. $\dfrac{1}{n^7} =$

Multiply and simplify:

13. $(y^7)(y^9)(y^{16}) =$
14. $(t^0)(t^0)(t^0) =$
15. $(x^5)(z^7)(x^{-12}) =$
16. $(a^{12})(a^{-4})(a^{-16}) =$

Divide and simplify:

17. $\dfrac{x^8}{x^2} =$
18. $\dfrac{x^{12}}{x^{12}} =$
19. $\dfrac{z^8}{z^{-7}} =$
20. $\dfrac{p^{-9}}{p^6} =$

Simplify:

21. $(z^2)^3 =$
22. $(p^5)^{-3} =$
23. $(z^{12})^{1/12} =$
24. $(z^2)^{1/2} =$
25. $(2p^2q^3)^3 =$
26. $2(3x^3y^4z^5)^2 =$
27. $\left(\dfrac{a^4}{3}\right)^3 =$
28. $\left(\dfrac{a^{-5}}{2}\right)^4 =$

EVALUATING POLYNOMIALS

Finding the value of a polynomial after replacing the variable with a number is called *evaluating* a polynomial. Here are some examples:

Example 18.1. Evaluating Polynomials

Evaluate the polynomial when x = 2

1. $4x + 5 = 4(2) + 5 = 8 + 5 = 13$
2. $3x^2 - 4x + 4 = 3(2)^2 - 4(2) + 4$
   $= 3(4) - 8 + 4 = 12 - 8 + 4 = 8$

Evaluate the polynomial when x = -2

3. $-8 - x^3 = -8 - (-2)^3 = -8 - (-8)$
   $= -8 + 8 = 0$
4. $-x^2 - 3x + 4 = -(-2)^2 - 3(-2) + 4$
   $= -4 + 6 + 4 = 6$

Like Terms

When terms have the same variable and those variables are raised to the same power, they are called *like terms*. Identify the like terms in the polynomials below.

Example 18.2. Collecting Like Terms

Polynomial	Like terms
1. $4x^3 + 5x - 4x^2 + 5x^3 + 3x^2$	$4x^3$ and $5x^3$ are like terms. $-4x^2$ and $3x^2$ are like terms.
2. $6 - 5a^3 - 8 - a - 5a$	6 and -8 are like terms. $-a$ and $-5a$ are like terms.

Descending and Ascending Order of Polynomials

When the exponent of terms in a polynomial decreases from left to right, the polynomial is in *descending order*. For example, the polynomial, $3x^6 - 5x^5 + 2x^3 - x + 2$, is said to be arranged in descending order because the term with the highest exponent is first, followed by the next largest exponent, and so on.

Conversely, when the exponents of terms in a polynomial increase from left to right, the polynomial is in *ascending order*. For example, the polynomial, $2 - x + 2x^3 - 5x^5 + 3x^6$, is arranged in ascending order because the term with the lowest exponent is first, followed by the next larger exponent, and so on. The term with the highest value of the exponent is last.

Degree of a Term and Polynomial

The *degree of a term* is the exponent of the term's variable. The *degree of a polynomial* in one variable is the *largest number* among the exponent(s) of terms. Here are examples of these concepts:

Example 18.3. Degree of a Term and Polynomial

Polynomial	Like terms
1. $-5x^6$	The degree of the term is 6.
2. $4x^3 - 4x^2 + 5x$	The degree of the polynomial is 3 because it is the largest exponent.
3. $6a^8 - 5a^3 - 8 - a^{11} - 5a$	The degree of the polynomial is 11 because it is the largest exponent.

Exercise Set 18.2

Evaluate the polynomial when $p = 3$ and when $p = -2$.

1. $-8p + 1$
2. $-3p^2 + p - 9$

Identify the like terms in the polynomials.

3. $2 - 4x + 3x^3$
4. $4x^3 + 5x - 6$

Collect like terms.

5. $5x^3 + 7x^3 + 3$
7. $3a^4 - 5a^3 - 6a^3 - 7a^4 + 2$

6. $5x^4 + 6x - 4x^4 - 7x$
8. $4p^6 + 3p^3 + 2p^3 - 5p^3 - 4p^6 + 3p + 6$

Arrange each polynomial first in *descending* and then in *ascending* order.

9. $x^5 + x + 6x^4 + 2x^2 + 5$
10. $11p - 7 + 7p^3 - 6p^4 + 3p^5$

Collect like terms and then arrange the polynomial in *descending* order.

11. $4x^4 - 6x^6 - 3x^4 + 7x^6$
12. $-2y - 2y - 2y + 2y^3 - 3y^3 - 4y^3$

Identify the degree of *each term* and the degree of the *polynomial*.

13. $3x^2 - 6x + 3$ 14. $2x^2 - 4x + 2x^6 - 11x^4$

Classify the polynomial as a monomial, binomial, or trinomial.

15. $x^2 - 25$ 17. 5

16. $60x$ 18. $6x^2 + 14x + 13$

Addition and Subtraction of Polynomials

Addition and subtraction of polynomials basically involves addition and subtraction of the coefficients of like terms. For example:

Example 18.4. Collecting Like Terms of Polynomials Separated in Parentheses

Add/Subtract Polynomial	Explanation
1. $(-4x^3 + 2x) + (5x^3 + 3)$ $= -4x^3 + 2x + 5x^3 + 3$ $= x^3 + 2x + 3$	Remove the parentheses first. Then collect like terms; adding the coefficient of the x^3 term.
2. $(-4x^3 + 2x) - (5x^3 + 3)$ $= -4x^3 + 2x - 5x^3 - 3$ $= -4x^3 - 5x^3 + 2x - 3$ $= -9x^3 + 2x - 3$	Remove the parentheses first and changing the sign of each term of the polynomial preceded by a negative sign. Collect terms with like signs and add the coefficient of the x^3 term.

Multiplying Polynomials

For the most part, multiplication of polynomials involves use of the distributive, associative, and commutative laws.

- For multiplication of a monomial and any polynomial, use the distributive law and multiply each term of the polynomial by the monomial.
- For multiplication of two polynomials, students need to use the distributive law more than once.
- Very often, students need to use a special technique for multiplying two binomials called FOIL. The abbreviation involves the multiplication of the *first* term in both binomials, then finding the product of *outside* terms, followed by the product of *inside* terms, and finally the multiplication of the *last* terms in both binomials. Some examples are provided next.

Example 18.5. Multiplying Polynomials

Products of Polynomial	Explanation
1. $3x(6x + 4)$ $= (3x)(6x) + (3x)(4)$ $= 18x^2 + 12x$	This is a product of a monomial and a binomial. First, multiply the monomial with each term of the binomial using the distributive law. Then evaluate each product.
2. $-4x(3x^5 + 3x^2 - 5x - 6)$ $= (-4x)(3x^5) + (-4x)(3x^2)$ $+ (-4x)(-5x) + (-4x)(-6)$ $= -12x^6 - 12x^3 + 20x^2 + 24x$	This is a product of a monomial and a polynomial with four terms. First, multiply the monomial with each term of the polynomial using the distributive law. Then evaluate each product using the product rule of terms with the same base.
3. $(x + 3)(x + 4)$ $= x(x + 4) + 3(x + 4)$ $= (x)(x) + (x)(4) + (3)(x) + (3)(4)$ $= x^2 + 4x + 3x + 12$ $= x^2 + 7x + 12$	Use the distributive law Using the distributive law on each term Multiply the monomials Collect like terms
4. $(x + 4)(x + 8)$ $= (x)(x) + (8)(x) + (4)(x) + (4)(8)$ $= x^2 + 8x + 4x + 32$ $= x^2 + 12x + 32$	Use the FOIL method Multiply the monomials Collect the like terms

Multiplying Sums and Differences of Two Terms

The product of the sum and the difference of the same two terms is the square of the first term minus the square of the second term. In equation form:

$$(A + B)(A - B) = A^2 - B^2$$ —the difference of two perfect squares.

As an example: $(z + 3)(z - 3) = z^2 - 9$.

Your students need to memorize this rule in both words and symbols. Review it with your students constantly, and make use of a "toolbox" (a review template written on one side of the board) to help students remember it and to emphasize its importance.

Division of Polynomials

Our focus here is on the division of a polynomial by a monomial. In this case, we use two rules:

- First, we use the quotient rule when bases are the same (subtract the exponents).
- Second, we divide the coefficients in the numerator and denominator.

For example:

Example 18.6. Dividing Polynomials

Division of Polynomials	Explanation
1. $\dfrac{18x^5}{3x^2} = \dfrac{18}{3}x^{5-2} = 6x^3$	First, divide the coefficients in the numerator and denominator. Second, subtract the exponents because the bases are the same.
2. $\dfrac{12x^9 + 18x^{11}}{3x^4} = \dfrac{12}{3}x^{9-4} + \dfrac{18}{3}x^{11-4}$ $= 4x^5 + 6x^7$	First, divide the coefficients in the numerator and denominator. Second, subtract the exponents because the bases are the same.

Squaring Binomials

The square of a sum (or a difference) of two terms is the square of the first term, *plus* (or *minus*) twice the product of the two terms, plus the square of the last term—that is:

$$(A + B)^2 = A^2 + 2AB + B^2$$
$$(A - B)^2 = A^2 - 2AB + B^2$$

Examples: $(y + 4)^2 = y^2 + 8y + 16$ and $(y - 4)^2 = y^2 - 8y + 16$.

Exercise set 18.3

Add:

1. $(3x + 2) + (-5x + 4) =$
2. $(x^2 - 11) + (x^2 + 11) =$
3. $(3x^3 - 13) - (3x^3 + 13) =$
4. $(x^4 + 2x^2) - (-2x^4 - 2x^2 + 5) =$

Simplify:

5. $-(x^3 - 4x^2) =$

6. $-(3x^4 + 4x^3 - 2x^2 - 5) =$

Subtract:

7. $(4x + 3) - (-5x + 4) =$

8. $(x^4 + x^3) - (-2x^4 - x^3) =$

Multiply:

9. $(8x^3)(2x^6)(2x^9) =$

10. $-4x(3x - 7) =$

11. $(x - 8)(x - 7) =$

12. $(x + 6)(x - 6) =$

13. $(4x - 2)(3x + 3) =$

14. $(3m + 4)(4m + 3) =$

Multiply mentally:

15. $(x + 5)(x - 5) =$

16. $(x + 1)(x - 1) =$

17. $(5m - 3)(5m + 3) =$

18. $(x^2 + 1)(x^2 - 1) =$

19. $(x + 3)^2 =$

20. $(2x - 1)^2 =$

21. $(x^2 + 3)^2 =$

22. $(4 - 3x^2)^2 =$

23. $(-z^2 + 1)^2 =$

Divide:

24. $\dfrac{27x^3}{9x} =$

25. $\dfrac{-18x^{11}}{-6x^{-3}} =$

26. $\dfrac{27x^3 - 36x^5 + 18x^3 - 45x}{9x} =$

27. $\dfrac{18x^3 - 36x^5 + 24x^4 - 42x}{6x^{-2}} =$

SAMPLE EXIT QUESTIONS

1. Evaluate: $(-b)^2$ when $b = -11$
2. Evaluate: $(-b)^3$ when $b = -4$
3. Evaluate: $-(-b)^2$ when $b = -6$
4. Multiply mentally: $(x + 1)(x - 1)$
5. Simplify: $\dfrac{-27x^{11}}{-9x^{-3}}$
6. Simplify: $\dfrac{21x^3 - 35x^5 + 14x^3 - 42x}{7x^2}$

SAMPLE PRIMING HOMEWORK

1. What is a common factor in $4x$, $8x^3$, and $16x^5$?
2. Factor what is common in all of the terms: $12x^3 + 6x^4 - 3x^7$
3. Factor: $(x^2 - 64)$
4. Factor: $x^2 + 7x + 12$
5. Factor: $y^2 - 8x + 7$

POINTS TO REMEMBER AND REVIEW

- The square of a negative term is always positive.
- The cube of a negative term is always negative.
- An even power of a negative term is always positive; an odd power of a negative term is always negative.
- A denominator divides into each term of the numerator. Exert extreme caution canceling terms in the numerator and denominator when there are multiple terms in the numerator and only one term in the denominator. For example, in $\dfrac{5x + 8x^3 + 11x^5}{5x}$, many students will cancel $5x$ in both the numerator and denominator—a popular but terrible mistake.
- Powers multiply when a term has a power and it is raised to another power; for example, in $(x^3)^5$, the powers of 3 and 5 multiply, giving the answer x^{15}.

19
Factoring

Factoring a polynomial means expressing it as a product of two or more polynomials. Essentially, factoring is the opposite of taking the products of two polynomials. Factoring is fundamental to algebra and those science courses that use it on a continuous basis. To factor, your students need to master the concept of greatest common factor (GCF); without it, factoring is not possible.

The *greatest common factor*, or GCF, is the largest number or term that evenly divides two or more terms. For example, the GCF of 10 and 15 is 5 because both terms can be divided evenly by 5, and it is the largest factor that does. Remind your students that they will use the GCF most of the time when dealing with factoring.

The *least common multiple* (LCM), on the other hand, is the smallest number that all numbers will divide into evenly. For example, the LCM of 10 and 15 is 30 because both 10 and 15 divide evenly into 30. The LCM is also known as the least common denominator (LCD). Remind your students that LCM is most commonly used when dealing with *fractions*.

FINDING GCF AND LCM

Example 19.1. Finding GCF and LCM

Problems	GCF	LCM
1. 2, 3, 5	1	30
2. 5, 10, 15	5	30
3. 4, 8, 12	4	24
4. 10, 20, 100	10	100
5. x, x^2, x^3	x	x^3
6. $3y^2, 6y^4, 12y^6$	$3y^2$	$12y^6$
7. $x, x(x-1), x(x-2)$	x	$x(x-1)(x-2)$
8. $(x-1)(x+1), (x+1)^2$	$x+1$	$(x-1)(x+1)^2$

METHODS FOR FACTORING

Factoring When Terms Have a Common Factor

In factoring, we usually come across polynomials that have more than one term. Some examples are:

Example 19.2. Factoring Polynomials

Factoring a Polynomial	Explanation
1. $8x^3 + 16$ $= 8(x^3 + 2)$	Factor out the GCF, 8
2. $12x^3 + 20x^2$ $= 4x^2(3x + 5)$	Factor out the GCF, $4x^2$
3. $24x^4 + 12x^3 - 36x$ $= 12x(2x^3 + x^2 - 3)$	Factor out the GCF, $12x$

Factoring by Grouping: Four Terms

When your students are dealing with four terms in a polynomial, there is a possibility (but no guarantee) that they can factor this polynomial by using a method called *factoring by grouping*.

Examples of this method are:

Factoring

Example 19.3. Factoring Polynomials, continued …

Factoring Polynomial	Explanation
1. $x^2(x + 1) + 3(x + 1)$ $= (x + 1)(x^2 + 3)$	The binomial $(x + 1)$ is common to both terms. Factor out the GCF, $(x + 1)$
2. $x^3 + x^2 + 3x + 3$ $= x^2(x + 1) + 3(x + 1)$ $= (x + 1)(x^2 + 3)$	x^2 is a common factor between the first two terms, whereas, 3 is a common factor between the last two terms. Finally, $(x + 1)$ is a common factor in the second step.
3. $6x^3 - 9x^2 + 8x - 12$ $= 3x^2(2x - 3) + 4(2x - 3)$ $= (2x - 3)(3x^2 + 4)$	$3x^2$ is a common factor between the first two terms, whereas, 4 is a common factor between the last two terms. Finally, $(2x - 3)$ is a common factor in the second step.

FACTORING QUADRATIC TRINOMIALS

General Rules for Factoring $ax^2 + bx + c$, when $a = 1$

Quadratic here means a polynomial of degree 2. Factoring of such polynomials can be best illustrated by taking some specific examples. Let's start with the case of trinomials of the form $ax^2 + bx + c$, when $a = 1$.

Example 19.4. Factoring Polynomials, continued …

Factoring Polynomial	Explanation
1. $x^2 + 10x + 16$ $= (x + 8)(x + 2)$	Factor 16 into two numbers whose product is 16 and their sum is 10. The numbers 8 and 2, whose product is 16, satisfy these conditions. Note: the numbers 4 and 4, which are factors of 16 as well, <u>do not</u> satisfy the sum condition.
2. $x^2 - 10x + 16$ $= (x - 8)(x - 2)$	Factor 16 into two numbers whose product is 16 and their sum is −10. The numbers −8 and −2, whose product is 16, satisfy these conditions. In this case, however, their sum must be equal to −10.
3. $x^2 + 6x - 16$ $= (x + 8)(x - 2)$	Factor 16 into two numbers whose product is −16 and their sum is 6. The numbers 8 and −2, whose product is -16, satisfy these conditions.
4. $x^2 - 6x - 16$ $= (x - 8)(x + 2)$	Factor 16 into two numbers whose product is −16 and their sum is −6. The numbers −8 and 2, whose product is −16, satisfy these conditions.

General Rules for Factoring $ax^2 + bx + c$ when $a = 1$

1. First, arrange the polynomial in the descending order.
2. Use a trial-and-error process that looks for pairs of factors of c whose sum is b and their product is c.
3. If c is positive, then the signs of the factors are the same as the sign of b.
4. If c is negative, then one of the factors is negative and the larger of the two numbers has the sign of b.

General Rules for Factoring $ax^2 + bx + c$ when $a \neq 1$

Now let's look at factoring trinomials $ax^2 + bx + c$, when $a \neq 1$. Again, we illustrate this by providing three examples. The technique used here is called the *ac method* or the *grouping method*.

Factoring Trinomial Squares

Some trinomials are squares of binomials, and there are two general types:

$$A^2 + 2AB + B^2 = (A + B)^2$$

and

$$A^2 - 2AB + B^2 = (A - B)^2$$

Example 19.5. Factoring $ax^2 + bx + c$ when $a \neq 1$

Factoring Polynomial	Explanation
1 $3x^2 + 10x - 8$ $= 3x^2 + 12x - 2x - 8$ $= (3x^2 + 12x) - (2x + 8)$ $= 3x(x + 4) - 2(x + 4)$ $= (x + 4)(3x - 2)$	Multiply the coefficients a and c [3 * (−8) = −24). Try to factor the product ac [−24 = 12* (−2)] so that sum of the factors is equal to b (10 = 12 − 2). Split the middle term and write it as a sum using the factors of ac. Factor by grouping.
2 $3x^2 - 10x - 8$ $= 3x^2 - 12x + 2x - 8$ $= (x^2 - 12x) + (2x - 8)$ $= 3x(x - 4) + 2(x - 4)$ $= (x - 4)(3x + 2)$	Multiply the coefficients a and c [3 * (−8) = −24]. Try to factor the product ac [−24 = (−12)* 2)] so that sum of the factors is equal to b (10 = −12 + 2). Split the middle term, and write it as a sum using the factors of ac. Factor by grouping.
3 $12x^2 + 26x + 12$ $= 2(6x^2 + 13x + 6)$ $= 2(6x^2 + 9x + 4x + 6)$ $= 2[3x(2x + 3) + 2(x + 3)]$ $= 2(2x + 3)(3x + 2)$	Factor out a common number if any; 2 in this case. Multiply the coefficients a and c (6 * 6 = 36). Try to factor the product ac [36 = 9 * 4)], so that sum of the factors is equal to b (13 = 9 + 4). Split the middle term and write it as a sum using the factors of ac. Factor by grouping.

In these polynomials, the left-hand sides are the quadratic trinomials, and the right-hand sides are the squares of binomials. There are certain conditions that your students can work with to find out whether a given expression to be factored is a trinomial square.

Examples:

Conditions for Trinomial Squares

1. The two expressions A² and B² are positive perfect squares, i.e., when their coefficients are perfect squares and the powers of the variables are even.
2. There must be <u>no</u> minus sign before A² or B².
3. The middle term is either +2AB or –2AB.

Example 19.6. Factoring Trinomial Squares

Factoring Trinomial Squares	Explanation
1. $x^2 + 6x + 9$ 2. $= x^2 + (2)(3)x + 3^2$ $= (x + 3)^2$	This is an example of a trinomial square because the first and the last terms are perfect squares, and the middle term is the product 2AB ($6x = 2*x*3$).
3. $4x^2 - 16x + 16$ 4. $= (2x)^2 - (2)(2x)(4)x + 4^2$ $= (2x - 4)^2$	Again, this is an example of a trinomial square because the first term and its coefficient and the last term are perfect squares, and the middle term is the product –2AB [$-16x = 2*2x*(-4)$].
5. $9p^2 + 16 - 24p$ $= 9p^2 - 24p + 16$ $= (3p - 4)^2$	The polynomial is not in the descending order. First, we write the polynomial in the descending order. Second, we examine the first term and its coefficient and the last term and we find them to be perfect squares. In addition, the middle term is the product –2AB [$-24p = 2*3p*(-4)$].
6. $27m^2 + 36mn + 12n^2$ $= 3(9m^2 + 12mn + 4n^2)$ $= 3(3m + 2n)^2$	First, we notice that the polynomial has a common factor of 3. Second, we examine the first and the last terms, and their coefficients, and find them to be perfect squares. In addition, the middle term is the product 2AB ($12mn = 2*3m*2n$).

Factoring Differences of Squares

The general equation for factoring the difference of squares is of the form:

$$A^2 - B^2 = (A + B)(A - B)$$

For simpler cases, your students need to recognize that A^2 and B^2 are perfect squares or the power of the variable is an even power. That is, when we take the square root of A^2 and B^2, *either* we get an integer answer *or* the power of the variable is evenly divisible by 2. Here are some examples:

Example 19.7. Factoring Differences of Squares

Factoring Differences of Squares	Explanation
1. $x^2 - 9$ $= (x+3)(x-3)$	Because x has an even power and 9 is a perfect square of 3, we can use the general form for factoring. Here, x^2 is a perfect square of x and 9 is a perfect square of 3.
2. $4x^2 - 25$ $= (2x - 5)(2x + 5)$	Again, x has an even power, its coefficient is a perfect square, and 25 is a perfect square of 5 as well. Therefore, we can use the general form for factoring. Here, $4x^2$ is a perfect square of $2x$ and 25 is a perfect square of 5.
3. $9p^2 - 36q^2$ $= (3p - 6q)(3p + 6q)$	Here again, the powers of p and q are even and their coefficients are perfect squares. Therefore, we can use the general form for factoring. Here, $9p^2$ is a perfect square of 3p and $36q^2$ is a perfect square of 6q.
4. $49m^4 - 64n^6$ $= (7m^2 - 8n^3)(7m^2 + 8n^3)$	Once again, the powers of m and n are even and their coefficients are perfect squares. Therefore, we can use the general form for factoring. Here, also, $49m^4$ is a perfect square of $7m^2$ and $64n^6$ is a perfect square of $8n^3$.

SOLVING QUADRATIC EQUATIONS BY FACTORING

The general form of a quadratic equation is

$$ax^2 + bx + c = 0.$$

We can solve this equation by using *two* different techniques. But, before we attempt to solve this equation, you need to review the principle of zero products with your students.

The Principle of Zero Products

Before solving a quadratic equation, the concept of zero products must be mastered. What this concept says is that, if $ab = 0$, then *either $a = 0$* or $b = 0$, or *both a* and *b* are zero. Your students can solve quite a number of problems in the STEM disciplines using this principle. Some examples for the zero products principle include:

Example 19.8. Using Principle of Zero Products

Solve	Explanation
1. $7x = 0$ $x = 0$	$x = 0$ is the solution.
2. $x(x - 5) = 0$ either $x = 0$ or $x - 5 = 0$ meaning $x = 5$	There are two answers for x. Both answers of $x = 0$ and $x = 5$ will satisfy the original equation.
3. $(x + 4)(x - 3) = 0$ either $x + 4 = 0$ meaning $x = -4$ or $x - 3 = 0$ meaning $x = 3$	Again, there are two answers for *x*. Both answers of $x = -4$ and $x = 3$ will satisfy the original equation.

Now turning our attention back to solving quadratic equations, we consider the two techniques mentioned earlier.

Using Factoring for Solving Quadratic Equations

Examples of using factoring to solve quadratic equations of the form $ax^2 + bx + c = 0$ are shown below.

Example 19.9. Solving a Quadratic Equation Using Factoring

Solve	Explanation
1. $x^2 + 7x + 12 = 0$ $(x + 3)(x + 4) = 0$ Either $x + 3 = 0$ giving rise to $x = -3$ or $x + 4 = 0$ giving rise to $x = -4$	This is a quadratic equation. Factor the left hand side. Equate one of the factors at a time equal to zero and solve for x. There are two solutions, i.e., $x = -3$ and $x = -4$. They both satisfy the original equation.
2. $x^2 - 10x = -25$ $x^2 - 10x + 25 = 0$ $(x - 5)(x - 5) = 0$ either $x - 5 = 0$ giving rise to $x = 5$.	This equation is not in the standard form. Write the equation in the standard form. This is a quadratic equation. Factor the left hand side. Equate one of the factors at a time equal to zero and solve for x. There are two solutions. In this case, both solutions are the same, i.e., $x = 5$, and it satisfies the original equation.
3. $4x^2 = 25$ $4x^2 - 25 = 0$ $(2x - 5)(2x + 5) = 0$ either $2x - 5 = 0$ giving $x = \frac{5}{2}$ or $2x + 5 = 0$ giving $x = -\frac{5}{2}$	This equation is not in the standard form. Write the equation in the standard form with $c = 0$, this is a quadratic equation. Factor the left hand side. Equate one of the factors at a time equal to zero and solve for x. There are two solutions. In this case, the first solution is $x = \frac{5}{2}$ and the second solution is $x = \frac{5}{2}$. Both solutions satisfy the original equation.

Using the Quadratic Formula for Solving Quadratic Equations

Most people will use the factoring method for solving a quadratic equation if possible. Sometimes, however, we cannot factor certain trinomials in the form of a quadratic equation easily. In that case, the *quadratic formula* can provide a helping hand. The quadratic formula for solving a general equation of the type $ax^2 + bx + c = 0$ is:

$$x = \frac{-b \pm \sqrt{b^2 - 4ac}}{2a}$$

Let's look at some examples:

Factoring

Example 19.10. Solving a Quadratic Equation Using the Quadratic Formula

Solve	Explanation
1. $x^2 + 7x + 12 = 0$ $(x + 3)(x + 4) = 0$ $x = -3 \text{ or } x = -4$ Now using the quadratic formula: $x = \dfrac{-7 \pm \sqrt{7^2 - 4(1)(12)}}{2(1)}$	This is a quadratic equation. Factor the left hand side. The solution obtained in the previous set of examples. Now let us use the quadratic formula: In this case, $a = 1$, $b = 7$ and $c = 12$.
$x = \dfrac{-7 \pm \sqrt{49 - 48}}{2}$	Simplify
$x = \dfrac{-7 \pm \sqrt{1}}{2}$	Simplify
$x = \dfrac{-7 \pm 1}{2}$	Simplify Simplify Simplify
$x = \dfrac{8}{2} \text{ or } x = \dfrac{6}{2}$ $x = -4 \text{ or } x = -3$	The answers are the same as obtained by using the factoring method.
2. $2x^2 + 20x = -25$ $2x^2 + 20x + 25 = 0$ Using the quadratic formula $x = \dfrac{-20 \pm \sqrt{20^2 - 4(2)(25)}}{2(2)}$	This equation is not in the standard form. Write the equation in the standard form. This is a quadratic equation. In this case, $a = 2$, $b = 20$ and $c = 25$.
$x = \dfrac{-20 \pm \sqrt{400 - 200}}{4}$	Simplify
$x = \dfrac{-20 \pm \sqrt{200}}{4}$	Simplify
$x = \dfrac{-20 \pm 10\sqrt{2}}{4}$	Simplify Simplify
The final answers are $x = \dfrac{-10 + 5\sqrt{2}}{2}$ or	Simplify
$x = \dfrac{-10 - 5\sqrt{2}}{2}$	Simplify

Exercise Set 19.1

Find the GCF of the terms.

1. x^2, $-8x$
2. $3x^4$, x^2
3. $10x^2$, $-5x$, -20
4. $10x^4$, $-30x^2$

Factor using the GCF.

5. $x^2 - 8x$
6. $8x^4 - 32x^2$
7. $5x^5 + 15x^3$
8. $8x^5y^3 + 16x^3y^2 + 24xy$
9. $x^2(x + 5) + 2(x + 5)$
10. $4z^2(3z + 1) + 6(3z + 1)$
11. $m^3(7 - 4m) - 3(7 - 4m)$
12. $8a^3(3a - 8) - 2(3a - 8)$

Factor by grouping.

13. $x^3 + 4x^2 + 2x + 8$
14. $5x^3 - 5x^2 - 3x + 3$
15. $12p^3 - 16p^2 + 6p - 8$
16. $2y^3 - 8y^2 - 4y + 16$

Factor the trinomials.

17. $x^2 + 11x + 18$
18. $b^2 - 9b + 18$
19. $x^2 - 5x - 24$
20. $x^2 + 2x - 24$
21. $x^2 - 72 + x$
22. $3y^2 - 15y + 18$
23. $7 - 8p + p^2$
24. $z^2 - 8z - 9$

Factor using the *ac* method and grouping.

25. $2x^2 + 11x + 5$
26. $5x^2 + 4x - 12$
27. $4x^2 - 4x - 15$
28. $2x^2 + x - 1$

Factor completely.

29. $x^2 - 18x + 81$
30. $x^2 + 20x + 100$
31. $x^2 - 22x + 121$
32. $x^2 - 24x + 144$
33. $5y^4 + 10y^2 + 5$
34. $m^2 - 2m + 1$
35. $m^2 + 2m + 1$
36. $x^3 + 30x^2 + 225x$

Factor completely—differences of squares.

37. $z^2 - 4$
38. $x^2 - 1$
39. $1 - x^2$
40. $4x^2 - 25$
41. $36 - 16y^2$
42. $a^2 - b^2$
43. $100 - r^2$
44. $81x^3 - 25x$

Solve using principle of zero products.

45. $(x + 4)(x - 8) = 0$
46. $y(y - 6) = 0$
47. $3y = 0$
48. $(x - 31)(x + 123) = 0$

Solve the quadratic equations using factoring.

49. $x^2 + 7x + 6 = 0$
50. $x^2 = 100$
51. $x^2 - 5x = 0$
52. $x^2 - 8x + 15 = 0$
53. $x^2 = 8x - 16$
54. $11x^2 = 11x$

Solve the quadratic equations using the quadratic formula.

55. $x^2 + 9x + 20 = 0$
56. $2x^2 + 9x + 10 = 0$

SAMPLE EXIT QUESTIONS

Factor completely.

1. $3x^5y^3 + 12x^3y^2 + 15x^2y$
2. $9a^3(5a - 11) - 2(5a - 11)$
3. $4p^2 - 25q^2$
4. $5a^2 + 10ab + 5b^2$

SAMPLE PRIMING HOMEWORK

Simplify:

1. $\dfrac{z^4 - 36}{z^4 - 1} \cdot \dfrac{z^2 + 1}{z^2 + 6}$
2. $\dfrac{3}{4} + \dfrac{5}{7}$
3. $\dfrac{3}{x} + \dfrac{7}{x^2}$
4. $\dfrac{1}{z - 1} - \dfrac{1}{z - 1}$
5. $\dfrac{1}{z - 3} + \dfrac{1}{z - 3}$

POINTS TO REMEMBER AND REVIEW

- Although GCF is an abbreviation for greatest common factor, the answers usually are the coefficients that are the *lowest* in value and the terms with the *lowest* exponents. For example, the GCF of $3x^2 + 9x^5 + 15x^4$—where the lowest coefficient is 3, and the term with the lowest exponent is x^2—is $3x^2$.
- Recognize the differences of two squares, and use the general form of factoring: $a^2 - b^2 = (a - b)(a + b)$.
- Recognize the complete squares of the sum and difference of two terms: $a^2 + 2ab + b^2 = (a + b)^2$ and $a^2 - 2ab + b^2 = (a - b)^2$.
- Bring all terms on one side of the equal sign before factoring.

20

Rational, Complex Rational, and Radical Expressions

Becoming an Algebra Guru

Here's the chance for each student to become an "algebra guru." At this point, whether students like it or not, remind them that the national tests emphasize heavily the topics of rational, complex rational, and radical expressions in their testing. Also, let them know these topics are the backbone of the STEM disciplines. The miracle is that each student really can become an algebra guru with full mastery of what is covered in this chapter.

The topics of this chapter are quite involved and difficult, and they can be tiresome, too. However, if the students truly master them, they will be rewarded by their performances in future college-level courses, and they will have genuine confidence in their ability to succeed. Also, finding jobs on campus as tutors or as teaching assistants may even further reward them.

We start our discussions with rational expressions followed by complex rational expressions. The last topic of this chapter will deal with radical expressions. However, before we present too many details about these expressions, let us review some of the concepts that we will need for the chapter.

REVIEW OF THE PROPERTY OF 1 AND THE LCM

Remind your students that multiplying with and dividing by 1 does not change the answer. They need to recognize that all of the following simplify to 1.

$$\frac{1024}{1024} = 1, \quad \frac{x+11}{x+11} = 1, \quad \frac{5x^2-6}{5x^2-6} = 1, \quad \frac{-1}{-1} = 1$$

The least common multiple (LCM) of two or more algebraic expressions is an expression that can be divided by all of the expressions. Some examples are:

Example 20.1. Finding LCM of Algebraic Terms

Find LCM	Explanation
1. $12x$, $16y$, $8xyz$ $= 48xyz$	The LCM of the coefficients is 48, whereas the LCM of the variables is xyz. Therefore, the final answer for LCM is 48xyz.
2. $2x^2$, $6xy$, $18y^2$ $= 18x^2y^2$	Again, the LCM of the coefficients is 18, whereas the LCM of the variables is x^2y^2. Therefore, the final answer for LCM is $18\,x^2y^2$.
3. $x^2 + 5x + 6$ and $x^2 - 9$ Factoring: $x^2 + 5x + 6 = (x + 3)(x + 2)$ and $x^2 - 9 = (x + 3)(x - 3)$ LCM, therefore, is $= (x + 3)(x - 3)(x + 2)$ or $= (x^2 - 9)(x + 2)$	First, find the factors of these two polynomials. Factors of the first polynomial Factors of the second polynomial The final answer for the LCM is divisible by both polynomials.
4. $x^2 - 25$ and $3x - 15$ Factoring: $x^2 - 25 = (x - 5)(x + 5)$ $3x - 15 = 3(x - 5)$ LCM, therefore, is $= 3(x - 5)(x + 5)$	First, find the factors of these two polynomials. Factors of the first polynomial Factors of the second polynomial The final answer for the LCM is divisible by both polynomials.
$x^2 - 1$ and $x + 1$ Factoring $x^2 - 1 = (x - 1)(x + 1)$ LCM, therefore, is $= (x - 1)(x + 1) = x^2 - 1$	First, find the factors of the first polynomial. Factors of the first polynomial The final answer for the LCM is divisible by both polynomials.
$x^2 - 6x + 9$ and $(x - 3)$ Factoring will give $= (x - 3)(x - 3)$ Therefore, the LCM is $(x - 3)^2$	First, find the factors of the first polynomial. Factors of the first polynomial The final answer for the LCM is divisible by both polynomials.

Rational, Complex Rational, and Radical Expressions 257

Recognition of these examples and the LCM are critical to the solution of some of the problems in this chapter.

RATIONAL EXPRESSIONS AND EQUATIONS

The *rational numbers* are the *quotients of integers* in which the denominator is not zero; likewise, *rational expressions* are ratios or *quotients of polynomials*. Quotients like 3/4, −8/13, and 113/1 are rational numbers. However, the quotients like 6/7, $\dfrac{5}{3+x}$, and $\dfrac{y^2 + 3y - 10}{y + 5}$ are rational expressions. As you can see, some quotients are *both* rational numbers and rational expressions. For example, 3/4 is both a rational number and a rational expression. For this chapter, we assume that the denominator is not zero.

Simplifying Rational Expressions

Your students can use a number of ways to simplify rational expressions. Dividing a smaller number into a larger one, or just finding a common factor of the coefficients, and then simply canceling the common factor in the numerator and denominator will work. They will also end up using the product and quotient rules of exponents with the same bases. Let us take some examples to explain these solutions further:

Example 20.2. Simplifying Rational Expressions

Simplify	Explanation
1. $\dfrac{16x^2}{24x}$ $= \dfrac{2x}{3}$	A factor of 8 divides into both 16 $\dfrac{16}{8} = 2$ and 24 $\dfrac{24}{8} = 3$. Use the quotient rule and subtract the exponents $(2 - 1 = 1)$ as the final power of x. Remember, $x = x^1$, therefore, $\dfrac{x^2}{x} = \dfrac{x^2}{x^1} = x^{2-1} = x^1 = x$
2. $\dfrac{6a + 12}{8a + 16}$ $= \dfrac{6(a + 2)}{8(a + 2)}$ $= \dfrac{6}{8} = \dfrac{3}{4}$	The numerator and denominator are factorable polynomials. Factoring both numerator and denominator and cancelling terms, $\dfrac{a + 2}{a + 2} = 1$ Simplifying

Simplify	Explanation
3. $\dfrac{x^2 + 3x + 2}{x^2 - 1}$ $= \dfrac{(x+2)(x+1)}{(x+1)(x-1)}$ $= \dfrac{(x+2)}{(x-1)}$	The numerator and denominator are factored into two factors that are polynomials. Factoring both numerator and denominator and cancelling polynomials, $\dfrac{x+1}{x+1} = 1$.
4. $\dfrac{x-6}{6-x}$ $= \dfrac{x-6}{-(x-6)}$ $= -1$	Recognizing that $x - 6$ and $6 - x$ are opposites, that is, $(x - 6) = -(6 - x)$. Cancelling the same polynomial, $(x - 6)$, in the numerator and denominator.

Multiplying, Dividing, and Simplifying Rational Expressions

Multiplying rational expressions amounts to basically writing them as a numerator and a denominator and then simplifying. Dividing them involves changing the division sign to a multiplication sign and then taking the reciprocal of the divisor (the expression after the division sign). Let us illustrate this by examples.

Example 20.3. Multiplying and Dividing Rational Expressions

Simplify	Explanation
1. $\dfrac{3x^3}{4} \cdot \dfrac{2}{3x}$ $= \dfrac{3x^3 \cdot 2}{4 \cdot 3x}$ $= \dfrac{x^2}{2}$	Two rational expressions written as products Writing both of them as single numerator and denominator Final answer after canceling common factors and simplifying

2. $\dfrac{x^2+6x+9}{x^2-4} \cdot \dfrac{x+2}{x+3}$ Two rational expressions written as products

$= \dfrac{(x^2+6x+9)(x+2)}{(x^2-4)(x+3)}$ Writing both of them as single numerator and denominator

Factoring

$= \dfrac{(x+3)(x+3)(x+2)}{(x-2)(x+2)(x+3)}$ Final answer after cancelling common factors

$= \dfrac{(x+3)}{(x-2)}$

3. $\dfrac{x+1}{x^2-1} \div \dfrac{x+1}{x^2-2x+1}$ The two rational expressions are written as division.

$= \dfrac{x+1}{x^2-1} \cdot \dfrac{x^2-2x+1}{x+1}$ Changing the division sign to a multiplication sign and then taking the reciprocal of the divisor.

$= \dfrac{(x+1)(x^2-2x+1)}{(x^2-1)(x+1)}$ Writing as one numerator and denominator

Factoring numerator and denominator

$= \dfrac{(x+1)(x-1)(x-1)}{(x-1)(x+1)(x+1)}$

Final answer after cancelling common factors

$= \dfrac{(x-1)}{(x+1)}$

4. $\dfrac{3x^2-5xy-12y^2}{3xy+4y^2} \div (3y^2-xy)$ Two rational expressions written in the form of a division

$= \dfrac{3x^2-5xy-12y^2}{3xy+4y^2} \cdot \dfrac{1}{(3y^2-xy)}$ Changing the division sign to a multiplication sign and taking the reciprocal of the divisor

$= \dfrac{3x^2-5xy-12y^2}{(3xy+4y^2)(3y^2-xy)}$ Written as one numerator and one denominator

Factoring numerator and denominator

$= \dfrac{(x-3y)(3x+4y)}{y(3x+4y)(y)(3y-x)}$

Using the property $(x-3y) = -(3y-x)$

$= \dfrac{-(3y-x)(3x+4y)}{y(3x+4y)(y)(3y-x)}$

Final answer resulting from cancelling common terms and simplifying

$= \dfrac{1}{y^2}$

Adding, Subtracting, and Simplifying Rational Expressions

We add and subtract rational expressions just the way we did rational numbers. Again, let's illustrate this technique with examples.

Example 20.4. Adding and Subtracting Rational Expressions

Add or Subtract	Explanation
1. $\dfrac{3}{8x} + \dfrac{7}{12x^2}$ $= \dfrac{3(3x) + 7(2)}{24x^2}$ $= \dfrac{9x + 14}{24x^2}$	The LCM of denominators is $24x^2$. For each rational expression, divide the denominator into the LCM and multiply the answer by the numerator. Simplifying
2. $\dfrac{2x}{x^2 - 1} + \dfrac{1}{x^2 + x}$ $= \dfrac{2x}{(x-1)(x+1)} + \dfrac{1}{x(x+1)}$ $= \dfrac{2x \cdot x + (x-1)}{x(x-1)(x+1)}$ $= \dfrac{2x^2 + x - 1}{x(x-1)(x+1)}$ $= \dfrac{(2x-1)(x+1)}{x(x-1)(x+1)}$ $= \dfrac{(2x-1)}{x(x-1)}$	Two rational expressions to be added. Factoring the denominators. The LCM of denominators is $x(x-1)(x+1)$. For each rational expression, divide the denominator into the LCM and multiply the answer by the numerator. Simplifying. Factoring the numerator. Final result after cancelling common polynomial factors.
3. $\dfrac{1}{x^2} - \dfrac{1}{x+1}$ $= \dfrac{1(x+1) - 1(x^2)}{x^2(x+1)}$ $= \dfrac{x + 1 - x^2}{x^2(x+1)}$ $= \dfrac{-x^2 + x + 1}{x^2(x+1)}$	Two rational expressions to be subtracted. The LCM of denominators is $x^2(x+1)$. For each rational expression, divide the denominator into the LCM and multiply the answer by the numerator. Simplifying. Writing the polynomial numerator in descending order.

4. $\dfrac{x}{x-y} - \dfrac{x}{x+y}$

$= \dfrac{x(x+y) - x(x-y)}{(x-y)(x+y)}$

$= \dfrac{x^2 + xy - x^2 + xy}{(x-y)(x+y)}$

$= \dfrac{xy + xy}{(x-y)(x+y)}$

$= \dfrac{2xy}{(x-y)(x+y)}$

LCM of denominators is $(x-1)(x+1)$.
For each rational expression, divide the denominator into the LCM and multiply the answer by the numerator.

Collecting like terms in the numerator

Simplifying

Solving Rational Equations

We have been adding, subtracting, multiplying, dividing, and simplifying rational expressions thus far, but we could not solve them. This section deals with solving rational *equations*. There are three steps to the solution.

1. Find the LCM of all the denominators.
2. Multiply all the terms on both sides of the equation with the LCM to get rid of the denominators.
3. Solve the equation that we obtain in step 2.

Once again, remind your students that solving the problems in the following sections requires lots of patience, attention, and tender loving care. One mistake at any point can be deadly for accurate problem solving. We will now illustrate this careful problem solving with several examples.

Example 20.5. Solving Rational Equations

Solving Rational Equations	Explanation
1. $\dfrac{2}{3} + \dfrac{8}{6} = \dfrac{x}{9}$	The LCM is 18
$18 \cdot \left(\dfrac{2}{3} + \dfrac{8}{6}\right) = 18 \cdot \dfrac{x}{9}$	Multiplying both sides by 18, the LCM
$18 \cdot \dfrac{2}{3} + 18 \cdot \dfrac{8}{6} = 18 \cdot \dfrac{x}{9}$ $12 + 24 = 2x$	Using the distributive law for the left side and simplifying Dividing both sides by 2, the coefficient of x
$36 = 2x$ $\dfrac{36}{2} = \dfrac{2x}{2}$ $18 = x$ or $x = 18$	Solution Simplifying
2. $\dfrac{1}{x} = \dfrac{3}{4-x}$	The LCM is $x(4 - x)$
$x(4-x)\dfrac{1}{x} = x(4-x)\dfrac{3}{4-x}$ $(4 - x) = 3x$ $4 = 3x + x$	Multiplying both sides by $x(4 - x)$, the LCM, and simplifying Transferring $-x$ from the left side to the right side of the equal sign and changing the sign of the term, and then simplifying
$4 = 4x$ $\dfrac{4}{4} = \dfrac{4x}{4}$ $1 = x$ or $x = 1$	Dividing both sides by 4, the coefficient of x Solution
3. $\dfrac{5}{x} = \dfrac{7}{x} - \dfrac{1}{4}$	The LCM is $4x$
$4x \cdot \dfrac{5}{x} = 4x \cdot \dfrac{7}{x} - 4x \cdot \dfrac{1}{4}$	Multiplying both sides by $4x$, the LCM and simplifying Rearranging terms
$20 = 28 - x$ $x + 20 = 28$ $x = 28 - 20$ $x = 8$	Solution

4. $x + \dfrac{12}{x} = -7$ The LCM is x

$x \cdot x + x \cdot \dfrac{12}{x} = x(-7)$ Multiplying both sides by x, the LCM and simplifying

$x^2 + 12 = -7x$ Rearranging terms and writing in descending order and factoring

$x^2 + 7x + 12 = 0$

$(x + 4)(x + 3) = 0$

Either $x + 3 = 0$ giving   Using the principle of zero products Solution

$x = -3$ or

$x + 4 = 0$ giving $x = -4$

Exercise Set 20.1

Simplify.

1. $\dfrac{9x^4}{36x}$

2. $\dfrac{22y^5z^7}{11y^3y^4}$

3. $\dfrac{5x - 15}{5x}$

4. $\dfrac{z^2 - 25}{z^2 + z - 20}$

5. $\dfrac{18}{x^4} \cdot \dfrac{5x^3}{9}$

6. $\dfrac{x^2 + 3x - 10}{x^2 - 4x + 4} \cdot \dfrac{x - 2}{x + 5}$

7. $\dfrac{x^2 + 11x - 12}{x^2 - 1} \cdot \dfrac{x + 1}{x + 12}$

8. $\dfrac{z^4 - 25}{z^4 - 1} \cdot \dfrac{z^2 + 1}{z^2 + 5}$

Find the reciprocal.

9. $\dfrac{6}{x}$

10. $\dfrac{1}{b + c}$

11. $\dfrac{x^3}{2x^2 - 4}$

12. $y^2 - 8y + 9$

Divide and simplify.

13. $\dfrac{m}{5} \div \dfrac{m}{15}$

14. $\dfrac{x^2}{y} \div \dfrac{x^4}{y^3}$

15. $\dfrac{6x-6}{7} \div \dfrac{3(x-1)}{14}$

16. $(3x-3) \div 3(1-x)$

17. $\dfrac{z^2-9}{5y+15} \div \dfrac{z-3}{10}$

18. $\dfrac{p^2-1}{6p+6} \div \dfrac{2p^2-4p+2}{12p+12}$

Find the LCM.

19. $6x^3$, $18x^2$, $24x$
20. $2(y-5)$, $6(y-5)$
21. z, $(z-3)$, $(z-6)$
22. (z^2-1), $(z+1)$

Add, and simplify if possible.

23. $\dfrac{3}{x} + \dfrac{7}{x^2}$

24. $\dfrac{7}{ab^4} + \dfrac{5}{a^2 b}$

25. $\dfrac{5}{x-3} + \dfrac{5}{x+3}$

26. $\dfrac{5}{x-1} + \dfrac{3}{(x-1)^2}$

27. $\dfrac{5}{x-1} + \dfrac{3}{(x^2-1)}$

28. $\dfrac{y-5}{y^2-16} + \dfrac{y-5}{16-y^2}$

Subtract, and simplify if possible.

29. $\dfrac{z}{z-7} - \dfrac{7}{z-7}$

30. $\dfrac{4m+2n}{2m^2 n} - \dfrac{3m+5n}{mn^2}$

31. $\dfrac{3}{y+7} - \dfrac{2}{y-7}$

32. $\dfrac{9}{y^2-4} - \dfrac{5}{y+2}$

33. $\dfrac{3}{x^2-x-12} - \dfrac{2}{x^2-9}$

34. $\dfrac{b}{b-a} - \dfrac{b}{b+a}$

Solve for the unknown.

35. $x + \dfrac{7}{x} = -8$

36. $\dfrac{x}{5} - \dfrac{5}{x} = 0$

37. $\dfrac{3}{x+1} = \dfrac{1}{x-3}$

38. $\dfrac{7}{y} = \dfrac{6}{y-8}$

COMPLEX FRACTIONS

A *complex fraction* is an expression that has one or more rational expressions within the expression. These expressions are probably the most difficult to deal with, but the principles are simple. Again, remind your students that they need to show lots of patience and care in simplifying these expressions. Some examples are:

$$\dfrac{1 + \dfrac{5}{x}}{7}, \quad \dfrac{\dfrac{x+y}{4}}{\dfrac{3x}{x-y}}, \quad \dfrac{\dfrac{1}{5} + \dfrac{2}{8}}{\dfrac{3}{y} - \dfrac{y}{x}}.$$

There are at least two methods for solving such complex rational expressions. We will describe the technique of using one of them that appears to be simpler and less error prone.

Procedure for Simplifying Complex Rational Expressions

1. Simplify the numerator to get a <u>single</u> rational expression.
2. Simplify the denominator to get a <u>single</u> rational expression.
3. Write the simplified expressions with the division (÷) sign.
4. Change the division sign to a multiplication sign and write the divisor as its reciprocal.

We now provide some examples to illustrate this procedure.

Example 20.6. Solving Complex Rational Expressions

Solving Rational Equations	Explanation
1. $\dfrac{\dfrac{3}{x} + \dfrac{1}{3x}}{\dfrac{2}{3x} - \dfrac{1}{4x}}$	A complex rational expression with a LCM of $3x$ for the numerator and $12x$ for the denominator
	Simplifying expressions
$= \dfrac{\dfrac{3(3) + 1(1)}{3x}}{\dfrac{2(4) - 1(3)}{12x}}$	Simplifying expressions
	Writing as division of two rationals
	Changing the division sign to a multiplication sign and taking the reciprocal of the divisor (the expression following the division sign)
$= \dfrac{\dfrac{10}{3x}}{\dfrac{5}{12x}}$	Simplifying
$= \dfrac{10}{3x} \div \dfrac{5}{12x}$	Cancelling and simplifying
$= \dfrac{10}{3x} \cdot \dfrac{12x}{5}$	
$= \dfrac{2}{1} \cdot \dfrac{4}{1}$	
$= 8$	
2. $\dfrac{1 - \dfrac{1}{x}}{1 - \dfrac{1}{x^2}}$	A complex rational fraction with a LCM of x for the numerator and x^2 for the denominator
	Simplifying expressions
$= \dfrac{\dfrac{x \cdot (1) - 1}{x}}{\dfrac{x^2 \cdot (1) - 1}{x^2}}$	Simplifying expressions
$= \dfrac{\dfrac{x - 1}{x}}{\dfrac{x^2 - 1}{x^2}}$	
$= \dfrac{x - 1}{x} \div \dfrac{x^2 - 1}{x^2}$	

Rational, Complex Rational, and Radical Expressions

Example 20.10. Multiplying and Simplifying Radical Expressions

Simplify	Explanation
1. $\sqrt{3}\sqrt{8} = \sqrt{3 \cdot 8} = \sqrt{24}$ $= \sqrt{4 \cdot 6} = \sqrt{4} \cdot \sqrt{6}$ $= 2\sqrt{6}$	First, write the products of two radical expressions into one expression. Second, multiply the two radicands to give 24, and then factor 24 into 4 and 6. The square root of 4 is 2.
2. $\sqrt{4x^4}\sqrt{8x^5} = \sqrt{4x^4 \cdot 8x^5} = \sqrt{32 x^9}$ $= \sqrt{16 \cdot 2 \cdot x^8 \cdot x^1} = \sqrt{16 x^8} \cdot \sqrt{2x^1}$ $= 4x^4\sqrt{2x}$	First, write as a single product and then multiply. Second, factor 32 into 16 and 2. Split the odd power into a largest possible even power (8) and an odd power (1). The square root of 16 is 4, and the square root of x^8 is x^4.
3. $\sqrt{20 y^2 x^3} \cdot \sqrt{10 y^9 x^6} = \sqrt{200 x^9 \cdot y^{11}}$ $= \sqrt{100 \cdot 2 \cdot x^8 \cdot x^1 \cdot y^{10} \cdot y^1}$ $= 10 x^4 y^5 \sqrt{2xy}$	First, write as a single product and then multiply. Second, factor 200 into 10 and 2. Split the odd power into a largest possible even power and an odd power. The square root of 100 is 10; the square root of x^8 is x^4, and that of y^{10} is y^5.

Dividing Radical Expressions

The principle to be employed here is that *the quotient of two square roots is the square root of the quotient of the radicands*, and vice versa. In algebraic terms,

$$\frac{\sqrt{A}}{\sqrt{B}} = \sqrt{\frac{A}{B}} \text{ and } \sqrt{\frac{C}{D}} = \frac{\sqrt{C}}{\sqrt{D}},$$

where A, B, C, and D are all positive. Examples:

Example 20.11. Dividing Radical Equations

Simplify	Explanation
1. $\dfrac{\sqrt{36}}{\sqrt{4}} = \sqrt{\dfrac{36}{4}} = \sqrt{9} = 3$	First, combine the two radicands into one radicand. Second, simplify the resulting radicand. The square root of 9 is, then, 3.
2. $\sqrt{\dfrac{64}{z^2}} = \dfrac{8}{z}$	The square root of 64 is 8; whereas the square root of z^2 is z.
3. $\dfrac{\sqrt{48x^9}}{\sqrt{4x^3}} = \sqrt{\dfrac{48x^9}{4x^3}} = \sqrt{16x^{9-3}}$ $= \sqrt{16x^6} = 4x^3$	First, combine the two radicands into one radicand. Second, simplify the resulting radicand. The square root of 16 is, then, 4; whereas the square rot of x^6 is x^3.
4. $\dfrac{\sqrt{200x^3}}{\sqrt{2x^{11}}} = \sqrt{\dfrac{200x^3}{2x^{11}}} = \sqrt{\dfrac{100}{x^{11-3}}}$ $= \sqrt{\dfrac{100}{x^8}} = \dfrac{10}{x^4}$	First, combine the two radicands into one radicand. Second, simplify the resulting radicand. The square root of 100 is, then, 10; whereas the square rot of x^8 is x^4. Note: Keep the power of the variable positive if you can.

Addition and Subtraction of Radical Expressions

Addition and/or subtraction of radical expressions is no different from those of expressions discussed in earlier chapters. Here again, we use the distributive laws and collect like terms. Let us illustrate the process with examples.

Example 20.12. Addition and Subtraction of Radical Expressions

Simplify	Explanation
1. $4\sqrt{7} + 5\sqrt{7} = 9\sqrt{7}$	Here, we <u>add</u> the coefficients of similar terms like $\sqrt{7}$.
2. $9\sqrt{6} - 2\sqrt{6} = 7\sqrt{6}$	Here, we <u>subtract</u> the coefficients of similar terms like $\sqrt{6}$.
3. $5\sqrt{7} + 2\sqrt{7} + 3\sqrt{7} - 4\sqrt{11} - 5\sqrt{11}$ $= 10\sqrt{7} - 9\sqrt{11}$	Here, we <u>add</u> the coefficients of similar terms like $\sqrt{7}$ and $\sqrt{11}$.
4. $(\sqrt{4} - \sqrt{x})(\sqrt{4} + \sqrt{x})$ $= [(\sqrt{4})^2] - [(\sqrt{x})^2]$ $= (4 - x)$	Here we use the property: $(A + B)(A - B) = A^2 - B^2$

Solving Radical Equations

A radical equation is not that much different from those we have solved in previous chapters. The only difference is that it contains radical expressions, that is, terms under the radical (square root sign).

Although solving radical equations is quite cumbersome, its usage in STEM courses is widespread, and the national tests are famous for including questions on it. Solving radical equations employs most of the principles and procedures that we have used previously.

To solve radical equations, we must first convert them into equations *without* radicals. Isolating the radical by placing it either on one side of the equal sign or on both sides of it and then squaring both sides of the equation will accomplish this conversion. *Caution: your students should consider checking the answers by substitution if they can.* Let us illustrate this by examples.

Example 20.13. Solving Radical Equations

Solve	Explanation
1. $\sqrt{3x} - 5 = 1$ $\sqrt{3x} = 1 + 5$ $\sqrt{3x} = 6$ $(\sqrt{3x})^2 = (6)^2$ $3x = 36$ $\dfrac{3x}{3} = \dfrac{36}{3} = 12$	A radical equation Transfer −5 to the right side of he equation to obtain the radical expression by itself on side of the equation. Simplifying Squaring both sides to eliminate the radical sign Solve for x by dividing both sides of the equation by the coefficient of x.
2. $2\sqrt{x-2} = \sqrt{x+0}$ $(2\sqrt{x-2})^2 = (\sqrt{x+0})^2$ $4(x-2) = (x+10)$ $4x - 8 = x + 10$ $4x - x = 10 + 8$ $3x = 18$ $x = 6$	Radical on both sides of the equal sign Squaring both sides Using the distributive property Using transposition Dividing both sides by the coefficient of x
3. $x - 5 = \sqrt{x+1}$ $(x-5)^2 = (\sqrt{x+1})^2$ $x^2 - 10x + 25 = x + 1$ $x^2 - 10x + 25 - x - 1 = 0$ $x^2 - 11x + 24 = 0$ $(x-8)(x-3) = 0$ $x = 8$ or $x = 3$	Only one side has the radical Squaring both sides Transposing Simplifying Factoring Using the principle of zero products

Exercise Set 20.3

Simplify.

1. $\sqrt{x^4}$

2. $\sqrt{a^4 b^8}$

3. $\sqrt{(a-b)^6}$

4. $\sqrt{100\, b^4 c^{12}}$

5. $\sqrt{9x^6}$

6. $\sqrt{28\, x^8}$

7. $\sqrt{(b^2 + 2c^8)^6}$

8. $\sqrt{32\, x^9 y^{11}}$

9. $\dfrac{\sqrt{36b}}{\sqrt{9b}}$

10. $\dfrac{\sqrt{63\, y^3}}{\sqrt{7y}}$

11. $\sqrt{\dfrac{49}{x^2}}$

12. $\sqrt{\dfrac{x^2 y^2}{144}}$

13. $\dfrac{\sqrt{7 z^{11}}}{\sqrt{63\, y^3}}$

14. $\sqrt{\dfrac{125\, p^4}{5 p^{18}}}$

Add or subtract.

15. $5\sqrt{3} + 6\sqrt{3}$

16. $11\sqrt{7} - 4\sqrt{7}$

17. $3\sqrt{5} + 4\sqrt{5} + 2\sqrt{5} - 7\sqrt{11} - 4\sqrt{11}$

18. $\sqrt{45} + \sqrt{80}$

19. $\sqrt{27} - \sqrt{12}$

20. $3\sqrt{50} - 3\sqrt{2}$

Solve.

21. $\sqrt{x} = 11$

22. $\sqrt{y+4} = 5$

23. $3 + \sqrt{x-1} = 5$

24. $6 - 2\sqrt{3n} = 0$

25. $\sqrt{2x-7} = \sqrt{x+10}$

26. $x - 5 = \sqrt{x-3}$

SAMPLE EXIT QUESTIONS

Simplify.

1. $\dfrac{8}{x-7} + \dfrac{8}{x+7}$

2. $\dfrac{\dfrac{a}{b} + \dfrac{b}{a}}{\dfrac{5}{b} + \dfrac{5}{a}}$

3. $\sqrt{\dfrac{2500\,p^8}{p^{20}}}$

4. Solve for x: $x - 2 = \sqrt{x-2}$

POINTS TO REMEMBER AND REVIEW

- LCM is the most important concept in this chapter. Practice finding the LCM of terms containing simple and complex expressions.
- Simplify within the radical sign before attempting to take the square root of terms.
- In distributive multiplication, take care of the signs, the coefficients, and then the variable terms, *in that order*. Many students miss the complex problems of this chapter because they have a tendency to put in the negative sign after they have computed other terms—and often, they forget to do it.

Finale

Well, we've finally arrived at the end of the line.
We hope you and your students came out just fine
With all their gaps in math neatly filled,
As well as any math fears fully chilled.

We also hope that they and you
Were truly amazed at how much they knew,
And the teaching principles we proposed at the start
In your hands have been raised to an elegant art.

… And so, from us, adieu.

Index

activities for first three days of teaching, 27; illustrations of teaching techniques and content of program, 27–28
addition and subtraction of fractions, 116–117
addition of whole numbers, 59–62, 75
addition with positive and negative numbers, 105, 108, 109, 111–112, 114
angles and intersecting lines, 98–100, 102–103, 105; definitions, 98–99, 105

basic math operations with two mixed numerals, 224–225; addition and subtraction, 224; multiplication and division, 226–227

classroom management, 1, 20; clusters of skills for students, 2–3; cluster of skills for teachers, 3–4
commission, 184
complex fractions, 265–268; procedures for simplifying, 265–267
compressed programs, pre-program orientation, 24. See also pre-program orientation, day 1; pre-program orientation, day 2; pre-program orientation, day 3
conditions for using the mean, median, and mode, 84–85
constants and variables with a power raised to a power, 145–146
coordinate system, 200, 201–203, 214–217;
coordinates of a point, 202–203; x- and y-axes, 201–203
cross products/cross multiplication (advanced),166–169; rules and steps in solving, 167
cross products/cross multiplication (simple), 161, 163, 164–166,169; basic concepts in . . . ,164–166, 169, 172; importance in STEM disciplines
currency, decimals, fractions, and percentages, 47, 54

discount, 180
division of fractions, 118–119
division of whole numbers, 67–68,75–76

elementary equations, 150, 151, 156–161

equation of a straight line, 207–211, 217; examples of finding the slope and intercepts, 210; finding the equation of a straight line, 211–212
estimating, 50
exponential notation, 231–235, 242; basic review of exponents and integers as exponents, 232–234; definitions, 231–232, 234; rules, 242
expressions, 150, 151–155; combining like terms, 154; cumulative review, 154–155; definition, 151; evaluating algebraic expressions, 152–153

factoring exercises for polynomials, quadratic trinomials and quadratic equations, 252–253, 254
factoring polynomials, 242, 243–244, 254; definition, 243; GCF and LCM, 243–244; methods for factoring polynomials, 244–245
factoring quadratic trinomials, 245–248; definition, 245; general rules, 246; trinomial squares and differences of squares, 246–248, 254
factoring to solve quadratic equations, 249; principle of zero products, 249
five major teaching strategies, 1
fractional powers of constants and variables, 147–148
fractions, 115–121; improper, 115; proper, 115
frequency distributions into a . . . , 86–89; bar graph, 86; frequency polygon, 89; frequency table, 87–90, 92; ie chart, 87, 91

GAP assessment strategy skills, 15; Definitions and illustrations of . . . 15–18; exit questions, 15–16, 20; "good" errors, 15; priming homework, 16–17
GCF, 13; definition, 13; illustration, 13
general formulas for dealing with exponents—overlearning, 149

improper fractions as mixed numerals, 221; definition, 221; rules for changing back and forth, 222, 230
inductive teaching strategy 4, 20 ; analogical reasoning, skill clusters of, 5; anchoring, skill clusters of, 5; definition, 5; equal participation, skill clusters of, 7, 20; illustrations of skills in . . . , 5; skill clusters, 5–6; students' differential knowledge bases, skill clusters of, 6, 11–12, 20
integers, 56–58, 73–74; definition, 56; numbers into word names, 57–58, 73–74; word names into numbers, 57–58, 74

LCM, 13; definition, 13; illustration, 13–14
linear equations, 157; characteristics of, 157; collecting and transferring—examples, 157; collecting, transferring, and dividing—examples, 158–159, 161; dividing by the coefficient—examples, 157–158

mathematical operations involving exponents with like bases, 141–145; addition and subtraction, 143–145; division, 142–143; multiplication 141
measures of central tendency, 81–84, 93; mean or arithmetic mean, 82; median, 81–83, 92, 93; mode, 83, 92
mental math exercises, 46, 50–53
multiplication of fractions, 117–118
multiplication of whole numbers, 65–67, 74–76
multiplication with negative numbers, 110–113, 114; general rules, 110; simplifying terms inside and outside of parentheses, 111, 112, 113, 114
multiplication tables, 45–46, 52–53, 77–78, 104, 113

overlearning, 18; definition and skills, 18

parallel and perpendicular lines, 212–213; definition of . . . , 213; equations and explanations for . . . , 213

parallel organization and teaching, 13; definition, 13; illustrations, 6–8, 14, 47–48; organization, 13

percentage, 185

perimeter, area, and volume, 191–200; equations for . . . by geometric shape, 195; formulas for familiar shapes, 193–194

polynomials—evaluating, 236–241; addition and subtraction of . . . , 238; definition, 236; degree of a term and . . . , 237–238; descending and ascending order of . . . , 236; dividing . . . , 240; like terms, 236; multiplying . . . , 238–239

powers of 10, 121, 125, 132, 136, 140; addition and subtraction, 130–132; definitions, 128; division, 128; mathematical operations, 128–132; multiplication, 128; positive and negative exponents to be overlearned, 126–127; standard powers of 10, 126–127; zero as exponent, 125

pre-program orientation, day 1: finding gaps in basic skills, 28–30; finding more gaps in basic skills, 32; identifying individual students' knowledge gaps, 28–34; student independent class work I, (day 1) 30; student independent class work II (day 1); teacher-directed instruction I (day 1), 28; teacher-directed instruction II (day 1), 32

pre-program orientation, day 2: finding more gaps, 34; finding more sophisticated gaps, 36–37; identifying students' knowledge gaps, continued, 34–38; student independent class work I (day 2) example, 35–36; student independent class work II (day 2) example, 37–38; teacher-directed instruction I (day 2), 34; teacher-directed instruction II (day 2), 36–37

pre-program orientation, day 3: review of a basic math departmental final examination, 38; student independent class work I (day 3), 40–41; student independent class work II (day 3), 42; teacher-directed instruction I (day 3)—basic math review, 39–40; teacher-directed instruction II (day 3)—basic math review (continued) 41–42; relationship of math scores to college performance, 24

range, 85, 93

radical expressions and equations, 268–276; addition and subtraction of . . . , 272–273; dividing . . . , 271–272; multiplying and simplifying, 270–271, 276; simplifying by factoring, 269; simplifying square roots of exponentials, 270; solving . . . , 268–269, 273–276

ratio and proportion, 173–175, 177; definition, 173; equal ratios, 174; procedures for setting up and solving problems, 173–175; using cross products to determine if ratios are equal, 175

rational expressions and equations, 257–265, 273; adding, subtracting, and simplifying . . . , 260–261; definitions, 257; multiplying, dividing, and simplifying . . . , 258–259, 276; simplifying . . . , 257–258; solving rational equations, 261–263

reading tables, graphs, and charts, 86–91

reasons for doing without calculator, 45

review of elementary equations using coefficients, 177

review of fractions and mixed numerals, 228–229

review of the property of 1 and the LCM, 255–257, 276

review with addition, subtraction, multiplication, and division of fractions, 120–121
rules regarding scientific notation and exponents, 136

sales tax, 182
simple interest, 187
scientific notation, 121, 123, 132–135, 136, 140; reasons for. . . , 132; specific rules for. . . , 132–133, 136
slope and intercept of a line, 204–207; negative slope, 206–207; positive slope, 205–206; x-intercept, 207; y-intercept, 206
square roots of whole numbers, 70–72, 76, 79
squares of whole numbers, 69–70, 76
squaring binomials, 240
streamlining course or program content, 19; definition and skills, 19; "parallel" curriculum design, 19; pruning, 19–20
subtraction of whole numbers, 62–65, 75

training model, 24; trainer, 24
triangles, 95–98, 102–103, 104, 105; characteristics, 96, 105; manipulating the Pythagorean Theorem, 97; types, 95–96

units of measurement and conversion, 100–101, 103, 104; three most basic types, 100–101
universals of teaching, 1

variables and constants, 137–138; definitions, 137; mathematical operation indicators, 137–138

Zero (0) and one (1): Facts, 48, 54, 72–73, 76–77, 79

About the Authors

Daryao Khatri is president/CEO of TopTech, Inc., and professor of physics at the University of the District of Columbia.

Anne O. Hughes is retired professor of sociology and anthropology at the University of the District of Columbia.

CPSIA information can be obtained at www.ICGtesting.com
Printed in the USA
BVOW071711061111

275349BV00003B/2/P